WORKING SHEEP DOGS

WORKING SHEEP DOGS

A Practical Guide to Breeding, Training and Handling

TULLY WILLIAMS

LAND
LINKS

National Library of Australia Cataloguing-in-Publication entry

Williams, Tully.
Working sheep dogs : a practical guide to breeding, training and handling.

Bibliography.
Includes index.
ISBN 9780643093430.

1. Sheep dogs – Australia. 2. Sheep dogs – Breeding – Australia.
3. Sheep dogs – Training – Australia.
4. Sheep dogs – Handling – Australia. I. Title.

636.737

Published by and available from
Landlinks Press, an imprint of CSIRO Publishing
36 Gardiner Road, Clayton VIC 3168
Private Bag 10, Clayton South VIC 3169
Australia
Telephone: +61 3 9545 8400
Email: publishing.sales@csiro.au
Website: www.publish.csiro.au

Cover photos courtesy of the author – email: tw@campaspeworkingdogs.com; website: www.campaspeworkingdogs.com

Set in 10/13.5 Adobe Minion and Helvetica Neue
Cover and text design by James Kelly
Typeset by Desktop Concepts Pty Ltd, Melbourne
Printed by Ingram Lightning Source

Foreword

In this book Tully Williams has portrayed an incredible amount of knowledge backed by hands-on experience. The combination of sheep dog, stock and handler forms scenarios of huge variation, with no definitive lines. It is difficult to write about.

However, Tully has interwoven his ideas to build a rock-solid case for what makes the good, all-round 'utility' dog. The depth in which he describes the breeding, training and working of sheep dogs is astounding.

Like a true sheep dog Tully is unassumingly quiet and calm, but can hit hard 'to the back teeth' when required. Obviously he has a great passion for good dogs and stock work, and has undertaken a superb study of the subject – naturally.

Tully grew up in the country and as a young man chose to work on large station properties. This experience, along with the fortune of having a few good dogs (and the frustrations of many lesser ones), has given him a valuable and well-respected quality of dog and stock sense.

Tully has always aspired towards improving the working sheep dog, and is a Council member and Judge of the Victorian Working Sheep Dog Association.

In recommending this handbook (I wish it had been available years ago) to those who wish to take the breeding, training and working of sheep dogs seriously, I congratulate Tully on his fine achievement.

Rod Cavanagh
(Author of *Australian sheep dogs – training and handling*)

Contents

Acknowledgements

I consider myself very fortunate that the first 'real' sheep dog person I had much contact with was Rod Cavanagh (author of *Australian sheep dogs – training and handling*), when I was 12 years old. He gave me some very sound advice, which I have never forgotten, and more since, and also encouragement.

Even now, having had wide experience of many sheep dog people, I am fairly sure that I couldn't have had anyone better to get me started on the right track, and I value Rod's friendship extremely highly. Not only is he a very knowledgeable dog man, but more importantly he is a man of rare integrity and character. Monetary gain means nothing to him compared with doing the right thing.

Rod was the first person I approached to read through some drafts of this book, and he was kind enough to do so a number of times. I made various changes thanks to his comments.

Sid Cavanagh (Rod's uncle), now in his mid-80s and still training dogs and competing in trials, is another that I count as a good friend. I have spent many hours talking – and working – dogs with Sid over the years, and many of his sayings stick in my mind. Sid has been competing in sheep dog trials for nearly 40 years, and has a wealth of knowledge. There is a lot of the blood of dogs that Sid owned, and bred, behind the best dogs I have had in the past, and behind those I have now.

Thanks also especially to my parents. They not only put up with my collection of dogs while I was still living at home, attending high school, but helped in innumerable other ways of course, over the years. Dad has also always been very willing to help me relocate between jobs, with all of the dog paraphernalia that that entails, and to help in anything else besides (including computer problems); I know I only have to ask. He has taught me to think critically, and not just accept things at face value.

Mum will also drop everything to help in any way, and she spends a lot of time helping people daily in her job, and in many other ways – a wonderful person. She has always been very encouraging and supportive of myself, and my two brothers and one sister, in whatever we have done.

Mum and Dad have both also read through various drafts of this manuscript, and made various suggestions that have influenced the final result. I knew that if Dad couldn't pick holes in the logic, then hopefully it was pretty sound, and that Mum would soon sort out the grammar issues (particularly my use of apostrophes!) and keep me in line in other things besides.

There are also a few other people over the years whose friendship I value, and from whom I have learnt one thing or another, such as Vin Stapleton, a retired stockman and experienced and knowledgeable dog man; Pam Vickery, who is one of the keenest dog people I know, and from whom I have had various dogs over the years; Les Hall and Mark Flood who are both good to work with in the dog game; and the late Jack Hiscock, who died late last year, and whose coffin, and surrounding areas in the church, were draped with a very great number of blue sheep dog trial ribbons he had collected over the years. Jack bred many high-quality working dogs. I will miss sitting at the kitchen table at the 'Moorlands' property and talking dogs over a cup of tea with Jack and Joan. Their son Robert has also been kind enough to allow me to use his good dog Moorlands Tomie as a sire.

There have been others also – many thanks to all.

Introduction

The very best working sheep dogs, with their incredible natural instincts, are remarkable animals. They are born knowing how to work stock, with that knowledge housed in an athletic body of speed, stamina and agility. Then, once a competent handler harnesses that knowledge, they perform amazing feats in trying conditions, far from the eyes of most. Only a lucky few are privy to witnessing such magic, and working such dogs.

However, in contrast to this very small number of top dogs, most modern working dogs are very poor. My greatest motivation for writing this book (perhaps somewhat optimistic) is that I may be able to bring about a reversal of this state of affairs.

One important reason for this general lack of quality is poorly designed sheep dog trials, because this is where most of the breeding originates. The necessity for good practical dogs has, in the case of most serious breeders, been replaced, diluted or modified by a 'necessity' for trial dogs. Many people also believe that dogs successful at trials are good *working* dogs, but this is usually not the case. Dog trials need a complete rethink in light of these facts.

The secret – if the working breeds are to be improved – is to bring the two together (that is, trials and practical work) so that the type of dog that excels at one is the same as that which excels at the other. Only then will we see a dramatic improvement in the breeding.

Most present-day handlers and breeders have not been fortunate enough to own a top dog, and so they have nothing worthwhile to compare their own – or other's – dogs with. They don't realise the amazing feats that the best dogs are capable of, and nor do they realise how poor their own dogs really are.

Throughout this book I have tried to explain what a good dog is, how it should work and the ability it should possess, and how trials should be designed in order to encourage the breeding of such dogs.

With today's technology of easy travel, communication, artificial insemination and so forth, it should be much easier to breed good dogs than it was in the past, *provided that breeders have the necessary knowledge, and provided that they have the necessary motivation*. (Motivation is a role that sheep dog trials often fill, particularly when real practical necessity doesn't apply.)

Perhaps the following quotes will help to show the heights that working dogs can attain, and show by contrast how poor most dogs are. First, the well-known Australian historian, C.E.W. Bean, had this to say in *On the wool track*, written in 1910:

And if there is any one agent in that long process from the paddock to the loom who shifts it [wool] swifter than any other, and from more awkward corners, through longer, hotter, wearier, dustier distances, by more continual use of his quick body and almost quicker brain, it is the sheep-dog.

There have been plenty of dogs who would let their master go ahead and choose a camp and get sticks, while they followed with the mob half an hour behind – or half a day, for that matter; or would even go on half a day ahead.

But to see the dogs do their ordinary work in the mountains is quite enough, without any parlour tricks. Sheep have a way – inherited, they say, from those long-forgotten centuries when through the old-world valleys there hunted wolves, and it may be, even beasts, whose very likeness has vanished many thousands of years – of huddling near the hill-tops in wild country; as night falls they invariably make for high ground. The result is that, in those districts where the paddocks stand mostly on end, or where the ridges take the place of fences, the sheep constantly feed in corners and crevices which the musterer's horses could not even get sight of unless they were able to climb like mountain goats.

But where the man can't go the dog will. At a sign he is off through the scrub, leaving you waiting on your horse at the mouth of the valley. You see him for a second or two far up on the skyline working through the rocks like an ant – and then he is over and gone.

Later – it may be ten minutes, or twenty, or even half an hour – he comes back with the sheep. You may turn your horse and go slowly on. He'll bring them up to you all right; the chances are he will not have left one behind. In New Zealand, I believe, the dogs do similar work – probably they do so where-ever sheep are fed in mountain country.

But at any rate the Australian who did not recognise that his companion in the cold fogs of the mountains, and in the teeming dust-laden short-tempered summer of the plains, was a fellow producer of wool, would be a very graceless Australian indeed. (Bean 1910)

The following quote is from that famous Australian dog man, breeder of Border Collies, and author of *The canine king*, James L. Moore. James Moore was responsible for importing MacPherson's Moss in the 1920s, amongst others.

To those of my readers who are unfamiliar with the demands made upon the sheepdogs and the wild country amidst which they are called upon to perform their tasks, it will be interesting to describe at short length the general work of sheepdogs and sheepmen in such country as the Monaro, where wild, rocky declivities shelter sheep which have straggled from the main flock, and which at mustering-time must be brought down to the valleys, often through other mobs of sheep which must be left undisturbed. No greater test of intelligence, of reasoning and quick thinking could be devised.

Having spent the best part of my life in rough country like this, I can, as I speak, visualise the work of these canine geniuses.

Rising early in the morning at 5 a.m., after a quick breakfast four of us saddle up our horses and each of us moves away with our dogs frisking around us as we start off on a ten-mile ride. We are charged with the task of bringing down from the hills a mob of some 4000 sheep scattered in fifty hollows, hidden behind rocks and in declivities, and some perhaps feeding unseen on the furthest boundaries of a paddock some 10,000 acres in extent.

It is impossible to give specific orders to a dog when one is unaware oneself of the position and numbers of the flock to be mustered. Besides, in such country the dog is soon out of hearing and sight. It is here that the definite reasoning powers of the dog are seen at their best. The dog is 'cast off on the blind', to use the expression by which a sheepman admits his own impotence. At a whistle from his master the dog casts out in the knowledge that he is on a general mission, in which his own sense is alone to guide him. Perhaps on the horizon his keen eyes note the form of very perceptible animals grazing. His nose to the ground on the scent of the sheep, yet with a vision that takes in a sweep of the country over which he is working, he goes forward.

Very shortly he is left to his own responsibilities. He goes out by a system of casting and recasting until he is outside the boundary of the furthest straggler of the flock now lying between him and his master. Coming up with this outlying body, he stops. He knows the consequences of precipitate action. He does not attempt to upset his charges by a hasty approach, but stands calmly in a commanding position until such time as he is satisfied that the sheep closest to him have observed his presence. Not until then does he attempt to 'lift' his sheep. This is the term used to describe the action of the dog in getting his sheep first in hand.

Slowly but surely he moves them forward, and if his roving gaze lights on some hitherto unobserved sheep he needs no direction, but springs into action and encircles them, guiding them firmly into the main flock. Then, shepherding them carefully into the tracks down the rugged hills to the valley below, where his master is waiting with other sheep mustered elsewhere by a canine lieutenant as tireless and clever as he, he is prepared for the general onward move.

The sheep have now settled down, and with the other dogs he takes up his position behind the flock, and his clear, ringing bark forces the mob forward until the yards of the shearing shed are reached and the mob is safely yarded for the next day's shearing. (Moore 1929)

The following quote from Rod Cavanagh's more recent book, *Australian sheep dogs – training and handling*, gives a good description of the real sheep dog:

'Complete' – 'utility' sheep dogs are capable of working wild cattle one day and successfully competing in a stylish arena trial the next; from mustering on a station property, to withstanding the trampling and squashing that goes with yard, shed and trucking work; from being strong enough to control and force hostile rams backwards up a ramp, to having the patience to manoeuvre and nudge new born lambs with the nose. They are capable of any conceivable sheep dog job imaginable… (Cavanagh 1990)

Unfortunately such dogs have become an endangered species. In fact, I have no doubt that many people will read these descriptions and consider them to be nothing but tall stories, and that such ability never really existed.

But such abilities do exist, and dogs like those portrayed above have existed in greater numbers in the past.

I hope that by the end of this book you will see the truth of this, and recognise that, if breeders put their minds to it, more dogs with these abilities can exist in the future – dogs with real holding ability, cover and balance, enormous searching casts, natural mustering ability, steady but with great force, without fear, very calm and with great intelligence. Such dogs are eminently possible, and I hope that this book will go part way to helping breeders and handlers produce such animals, and to enjoy such ability.

I also hope that this book will give the general farmer and stockperson the knowledge necessary when buying and training dogs, and that they will demand much more from the breeders, who have let them down badly. Many trial breeders want their cake and to eat it too – they want to breed trial dogs mostly not suited to farmers, and then they want to sell them to farmers! And even breeders breeding farm dogs often have little idea of what a good dog is.

The author working stock on *Madness*

As far as training is concerned, a few simple points can greatly impact on a dog's effectiveness and its ease of training. One such point is the slight modification to the commanding system that can result in a large improvement in performance, with no real extra effort. I hope the notes on training will help both those very ambitious handlers wanting the highest calibre dog work, and also make things easier for those only wanting enough to get by. I outline methods for training various skills in the quickest, easiest, and most effective manner.

There is real magic in witnessing and working the truly good dog mustering in rough country, and in the partnership between man and animal when the animal plays such a large role; or in seeing the top dog handling wild sheep out in the hills among the rocks and the timber, but unfortunately such dogs are rare, and few people have been so lucky. My greatest hope is that this book will furnish keen dog people with enough knowledge to breed the real 'genius' back into the working dog.

Part 1
NATURAL ABILITY AND SELECTION

1

Natural ability

'Natural ability' is often discussed when sheep dogs are the topic. It simply means the dog's natural instincts as they relate to working stock.

Natural ability is the most important aspect of the working dog.

Just as a bitch knows how to chew the umbilical cord and nurse her pups without having to learn, or as a beaver knows how to build a dam, so too the good sheep dog *knows* how to work stock. It isn't (or shouldn't be) a learnt skill, but an inherited one.

So why should this ability be inherited, rather than learnt? There are two reasons.

- The natural worker requires much less training, and becomes a better dog more quickly and easily.
- Only the dog with the right instincts will handle stock well, *particularly in difficult situations.*

There are simple instincts, and there are complex instincts. For example, a beaver building a dam is an example of a complex instinct, involving various aspects over time. On the other hand most of the sheep dog's instincts are simple instincts, or reflexes. The dog simply responds to the sheep's movements or actions in a reflex manner, depending on which instincts it has inherited.

For instance, when a sheep is running away from it, one dog will run straight for the sheep and attempt to grab hold of it. However, the good dog will attempt to get around in front of the sheep to stop it – it will get wider out from the sheep as it heads it off and arc wide around in front of it. This is an instinctive, reflex action.

Yet when many people talk about natural ability, such as the more mechanical-style (and often successful) three-sheep trial handlers, or many UK trial handlers, they often simply mean the right make-up to best enable the dog to be turned into a robot, working completely under command. There is this type of so-called natural ability, and then there is *real* natural ability.

An example of a somewhat complex instinct in the sheep dog is the extraordinary mustering instinct, which has almost been lost. The ability to muster a big paddock of scattered stock, searching for them, putting some together and starting them on their way, then breaking back out for others and starting them on their way, and so on, until they are all together in a mob, then taking hold of the mob and working it where the dog wills, is instinct at its best. Not many people have seen such natural ability.

It must be realised that instinct is different from intelligence. You can have a dull dog with good instincts, or a very intelligent dog with poor instincts. The two are more or less separate, and many people confuse them. Just because a dog will muster a paddock on its own does not mean that it is highly intelligent – intelligence is more related to a dog's ability to learn things quickly. An intelligent dog can be taught a new command or manoeuvre very quickly, whereas the less intelligent dog requires more repetition.

Bob Ross receiving the trophy from Queen Elizabeth II, after winning the 1970 National Sheep Dog Trial at Canberra with Yulong Russ. Russ was an exceptional natural worker and influential sire. (Photo reproduced courtesy of *The Herald & Weekly Times* Pty Ltd)

However, the speed at which a dog learns is also affected by other aspects, such as its temperament and instincts. For example, a bold dog may not learn some things as quickly as a timid dog, only because it isn't as sensitive; yet it may be just as intelligent. The sensitive dog may seem more intelligent in that it is easier to influence, but it may be an illusion. And a dog with natural casting ability may appear to learn to cast more quickly, simply because its instincts guide it.

Instinct and age

A dog's instincts don't change with age – they may become evident with age, but they don't change. That is, some instincts may not be apparent until the dog matures a bit, such as a dog lifting its leg.

Instincts can show at different ages in different dogs, and the age at which they show is inherited. Some dogs' 'working instinct' (the keenness to work stock) surfaces very young, before weaning even. In others it may not surface until 10 months old, or even older in some cases. Personally I like to see it developed fairly strongly by about four months of age, otherwise you can waste too long waiting to see whether a pup is going to be any good or not.

And, despite things that have been said to the contrary, there is no other advantage in either late or early starters. You can have late starters that are over-keen and excitable, and you can have early starters the same. And you can have early or late starters that are calm and sensible. It all depends on the bloodline.

However, once a pup starts, I like to see it start properly. Some pups show some interest, but only in a half-hearted fashion; most pups of this type are 'weak' – fear is holding them back. Often they show interest while the sheep are moving, but then, if the sheep stop and look at them, they lose interest. As they gradually gain confidence they begin working more strongly. However, some weak pups can still start strongly, particularly if they have very strong instincts and a lot of 'eye'; while in contrast a gradual starter may be that way simply because it has only weak instincts.

So, although instinct surfaces at varying ages, once it has surfaced it doesn't alter. Its expression may alter through gains in confidence or training, but the instinct itself doesn't alter, and *the basic instinct will always surface when the dog is pushed to its limits*.

This means that a dog that doesn't have natural holding ability as a pup will never have real holding ability, even though you may teach it to keep out wide and so on. So don't make too many excuses for a pup's faults, because *a pup's faults will become an adult dog's faults*.

An ideal testing ground to push the best dogs to their limits

This is particularly true in practical work. If a dog is kept solely for trial work, or on small or easy properties, you may be able to hide its faults; however, if you push the dog to its limits in hard practical work they will always surface.

Basic premise for breeding working dogs

The basic premise for breeding sheep dogs is that you should never have to teach a dog *how* to work sheep. You should only have to teach it a system of commands that allows you to communicate your wishes. Everything else with regard to handling the stock should be inherited natural instinct.

The good dog should keep all the sheep together and never split them unless taught to do so; it should muster a paddock of scattered sheep; cast long distances with very little training; balance and hold and cover wild sheep; it shouldn't ring sheep on the draw; it should force strongly but cleanly when required, but not otherwise, and so on. This is the real sheep dog – the natural sheep dog.

You can breed for just about anything you set your heart on, provided that you understand how to go about it. Nearly every aspect of the working dog's behaviour is, or can be, inherited. For example, the place where the dog bites (if and when and how it does); the way it sits when riding on a motorbike; how many steps it takes before stopping when told; and so on. These things are all inherited (they can be modified by training) but you can breed for anything you like.

The trick is in knowing what you want, and knowing what makes a top dog.

Instinct is the most important factor of the working dog, because it determines how well the dog will handle stock, particularly in difficult situations or if left to work on its own resources, and how easy it will be to control and to work, and how much it will think about what it is doing.

Strength of instinct

A dog's instincts can vary from weak to very strong. For example, some dogs' heading instinct (particularly many three-sheep collies) is excessively strong and requires a lot of control to harness it, whereas a dog whose instinct is not so strong will be easier to control.

However, some breeders, in trying to breed easily controllable dogs (which is good, depending on how it is achieved), simply breed dogs with fairly weak instincts, that is, lacking real keenness. But, even though such dogs are easier to handle, they tend to lack much 'heart' (the desire to keep working in the face of just about any hardship).

It is possible, though, to breed controllable dogs with strong instincts and great heart. The clue lies in breeding *calm* dogs with the right instincts, as explained later.

The working traits

In the following chapters I have provided a reasonably comprehensive breakdown and analysis of the instincts and qualities that combine to make the working dog.

A knowledge of them allows a more accurate assessment of a dog's abilities, and hence more accurate selection for breeding.

If you are ever going to breed top dogs, then, rather than just thinking that a dog is 'a good dog', you *must* be able to break its work down into individual traits, so you know exactly *why* it is a good dog.

A clear understanding of the make-up of the good dog will help anyone to breed better dogs. Only then will you be able to develop a clear goal in your mind of exactly what a good dog is, and what you are trying to breed and select for, so that you aren't fighting against yourself (which can easily happen).

One vital point is that it is difficult to assess many of these characteristics when the dog is seen working quiet stock in a small area, or under tight control. You must see it tested in difficult situations, and left to work on its own resources.

For example, it is difficult to know how much 'mob cover' some dogs have, when they are working a few quiet sheep that plod about and stick together. Only when you see them handling a dozen wild, shorn wethers or similar, which are running and splitting and carrying on out in a paddock, can you get a good idea.

Always make a point of testing a dog beyond its limits. Only by seeing it pushed beyond its limits can you see its limitations.

The modern trend to work dog trials on quietened sheep is deplorable (and this includes not only three-sheep trials but also trials such as the national Kelpie utility trial). Not only do these trials greatly reduce the challenge and skill involved, but also

Pat Murphy's high quality, big casting, all-round Kelpie dog Paddy's Shadow (Karrawarra Boogles × Paddy's Loo). (Photo courtesy Pat Murphy.)

they don't provide the breeder with a sufficient insight into the dog's natural abilities. Many dogs look good on quiet sheep, if they are well trained. But only the high quality natural dog will look good on wild sheep (or wild cattle, for that matter). Sheep dog trial handlers with the better natural dogs sometimes comment that the only time they can get among the winners is when wild sheep are worked, because with quiet sheep the more mechanical dogs can score highly.

People often reminisce about the 'old-time Kelpie' and regret that such dogs are rarely seen today. But there is nothing mysterious about those old-time dogs; they were simply the product of experienced and knowledgeable stockmen selecting for a certain type of practical dog, without any thought of breeding to win trials. The same can be said about the 'old-time' Border Collie.

The reason those dogs aren't common now is because breeders aren't selecting for them; they don't understand what it was about those old-time dogs that made them the way they were. In particular, most breeders don't understand the roles that 'drive' and 'minimal activity' play, or what the 'right type of eye' is, or 'distance', or the way that a dog should 'force', or how important a calm, relaxed temperament is.

It is entirely possible to breed such outstanding 'old-fashioned' dogs from the dogs that are about today, *if breeders understand what makes a good dog.*

In the chapters that follow I have broken the dog's make-up down into a number of areas (which can be further broken down). In order to be any good a dog must be capable in all areas.

These areas are:

1 temperament and intelligence
2 steadiness and minimal activity
3 controlled force
4 holding ability
5 driving ability
6 cast and muster.

Some of the terms used may be new to some readers, and therefore I suggest you read through the following chapters on the working instincts once fairly quickly, just to get an overview of the various terms, and then study them more fully.

2

Temperament, intelligence, steadiness, minimal activity, and force

I have grouped these aspects into one chapter, because they are all heavily dependent on the dog's temperament. A dog with a poor temperament will be deficient in many of these areas.

Temperament

Temperament is at the top of the list, because it affects every area of a dog's work. I will repeat that again: *temperament affects every area of a dog's work*. The reason why many dogs are so poor is because not enough emphasis is placed on temperament.

Many people hope that they can breed a poor tempered dog (often weak, timid or tense) which they think shows a lot of ability (break, eye, cast, caution), to a better tempered dog (stronger and calmer), and get the ability of the first with the temperament and strength of the second.

This fails because you cannot separate ability from temperament in this way. The timid, soft, or tense dog's temperament is usually what causes it to work in the way that it does – wide, cautious, strong eyed. If you changed its temperament you would change its work simultaneously.

Work and temperament are inseparable.

This is why I have placed temperament at the top of the list. You cannot have a top dog with a poor temperament. It is simply not possible.

A good temperament

So, what is a good temperament? In two words, *calm* and *bold*. These are the main things to look for. You can see this in pups long before they go anywhere near sheep, and select accordingly. In the majority of litters being bred you could downgrade most pups on one or the other of these points – they will show either excitement, or some level of timidity or softness.

The type of calmness we should be looking for is seen in the dog that can readily be trained to lie down in one place while you work other dogs, even if there is a lot going on. I could lie Boho Misty (a collie bitch I had) down in the middle of a yard full of sheep, with other dogs barking and forcing and sheep going all over the place, jumping over her and around her, and she would lie there perfectly content until told otherwise. And yet she was very keen, with great heart.

A simple test is to pick up a young pup when it is about four or six weeks old (or older) and see how it reacts. The good pup will just hang there relaxed and unconcerned. Poorer pups will wrap their front legs around your hand and hang on for grim death, or tense themselves (particularly if you move them about a bit), while others will wriggle and squirm.

Other than that, simply look for any signs of a calm, relaxed nature.

As far as boldness is concerned, we look for pups that are tough and unafraid. A good test is to rattle a tin full of pebbles near them, and watch their reactions, or clap your hands or crack a stockwhip, or wave a bag about. Timid and soft pups will disappear into their kennel and hide, while bold pups will be more or less oblivious. I also like a pup that will look you calmly in the eye.

However, many dogs are timid and whip-shy (even though some of them may still be very keen with good heart).

Others have problems with sulky and soft temperaments, and the resultant lack of heart. Such dogs may be easier to control and influence, but they must be handled with kid gloves or you can put them off work entirely – they lose interest easily when under pressure. Such dogs are undesirable. (I have found this to be a problem with many, but not all, Kelpies, and I believe this is one reason why Border Collies predominate in three-sheep trials.)

For example, if things are going 'pear-shaped' as sometimes happens in stock work, such as a mob of young cattle splitting up all over the place in rough country, with stockmen (and you) yelling and roaring, and stockwhips cracking, such dogs that cannot take pressure in a

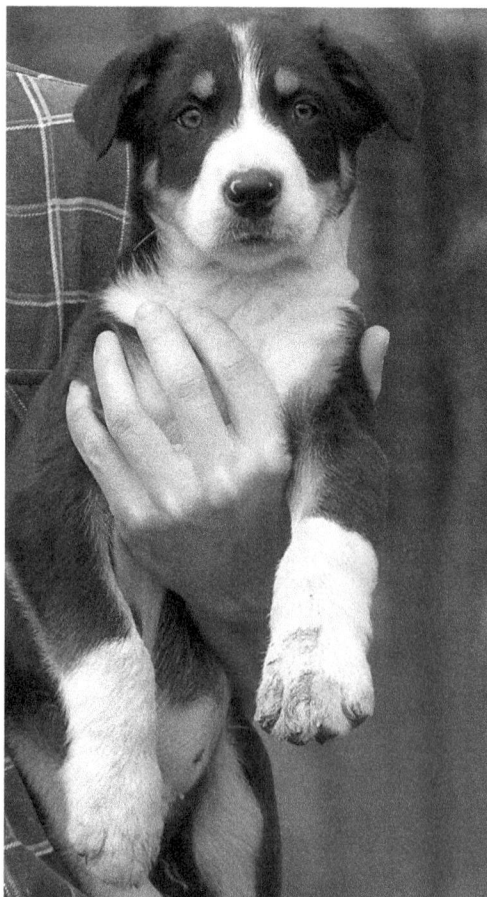

A 10-week-old Border Collie pup in-hand

training situation will also crumble under this type of pressure, and will leave you in the lurch right when you need them most.

This means that a dog that cannot take pressure in a training situation and still remain keen and working, with plenty of heart, is far from ideal. The best dogs not only have exceptional natural ability, but will also remain keen even when the trainer puts a lot of pressure on them to train them quickly.

So look for the calm, bold dog with plenty of heart. Steer clear of soft or timid ones, or half-hearted ones that have to be nursed along. As James Moore said, 'Above all things, never breed from a soft-tempered dog' (Moore 1929).

Intelligence

Intelligence is a valuable trait, but the amount of 'brains' a dog shows in its work is more dependent on its working instincts, and temperament, than on its intelligence alone. (This is explained fully in Chapter 4, 'The driving traits'.) For example, an excitable or busy dog, or one with too much 'eye' or too keen on the head, will not think about what it is doing no matter how intelligent it may be.

Even when just taking pups for a walk you can see how their temperament affects their ability to use their intelligence. If you climb through a fence, the calmer pups will generally find a way through after you, whereas those with poorer temperaments will often yelp and run backwards and forwards. *It isn't that they are necessarily less intelligent, but that their temperament doesn't allow them to USE their intelligence.*

The advantage of the intelligent dog is that it learns more quickly, and so develops into a better dog (with a good handler). However, it can also learn bad habits more quickly, and run rings around some handlers.

An example of the speed at which a really intelligent dog can form either good or bad habits involves a collie bitch I had called Boho Misty. When first teaching her to ride on the back of a motorbike she ran around the bike a couple of times (clockwise), because she didn't understand what I wanted, before finally jumping up. Whenever I asked her to jump up on the bike after that she would always do two clockwise circles of the bike before jumping up.

After a few months I became annoyed with this, and decided it was time to change things. So I parked the bike with its front wheel up against a fence to prevent her circling it, and asked her to jump up. She tried to get around the bike a few times, but eventually jumped on board. I made her do this a few times, then tried it out in the open and prevented her from circling the bike by voice, and from then on she was cured. Less intelligent dogs wouldn't have developed the bad habit in the first place, because they don't associate things so quickly.

Misty was also a real escape artist, as many intelligent dogs are. The only sure way of preventing her escaping was a dog chain, or a fully enclosed dog cage. One time she even got out of a corrugated iron shed with a concrete floor. She was locked in there while she was on heat, and one day I was out working stock when she turned up to help. When I got home I examined the shed. I found that she had ripped a hole in the corrugated iron on one wall and found hay piled against the other side of it, so had then tried another wall and by pulling a sheet of iron loose she had got out. But she had done it calmly and deliberately – she wasn't a dog that ever got excited or frustrated. She would never bark or become agitated if I was working another dog and left her on the chain, rather she would sit calmly on her tail and howl.

Anticipation

Anticipation is also a valuable trait to have in a dog. For example, when I let all the dogs off for a run around, and then prepare their feed, they mostly stand waiting at the door into the feed shed and then follow me around as I feed each dog. But a collie bitch I have called Cullens Kirribilli will make a beeline in the opposite direction for her kennel, and will wait there.

Her daughter, Campaspe Sue (three-and-a-half months old), is similar. She and her litter sister are running in a yard together, with another enclosed dog-run inside that yard. I have to feed them separately or the other pup gets all the feed. So I put one feed bowl down in the outer yard, then Sue's in the inner enclosure. As soon as I go to feed them, Sue runs away and waits in the inner enclosure.

This is a good sign, and suggests that Sue will have good anticipation when working. In fact I can see this when she is having a free run with a few sheep. If one breaks away she will break off it and race to get well ahead of it, to turn it back. She is anticipating where she needs to be, *and is no longer working the sheep directly*, but races for a spot she picks out up ahead.

Cullens Kirribilli

Campaspe Sue and litter mate holding sheep at three-and-a-half months

Anticipation when working is related to 'break' (discussed in the next chapter), but not all dogs with break have it. It is hard to explain the difference between a dog just breaking wide and one showing this 'anticipating', or 'repositioning' break, but there is a difference.

This type of dog will learn (if its instincts are right) to anticipate where sheep usually break around a hillside or head down a wrong road, and will be there before the sheep to prevent it, but will also be superior in less obvious ways (this ability is one aspect of 'casting break', described in Chapter 5).

For consistent levels of exceptional intelligence, the UK dogs leave any others for dead. They have been selected for their ability to learn quickly, due to the nature of their dog trials. However, I would always prefer a dog of great natural ability and only average intelligence, to a dog of lesser natural ability and superior intelligence, but it is best if you can find both.

Steadiness

Steadiness is a quality I rate very highly in the practical working dog. It is largely a result of the right temperament, that is, *calmness*. The good dog should walk or trot whenever possible, only increasing pace when needed.

The constantly fast dog only takes stock too quickly, doesn't think enough about what it is doing, is harder to control, and knocks itself about.

However, be careful not to confuse steadiness with 'doughiness'. The doughy dog is undesirable. Even though a steady dog should just poke along, it should do so in a *free moving* manner and have a fast turn of pace when required. A good analogy is to consider how honey flows (doughy), or how water flows (free moving), and relate that to how a dog works.

'Stickiness' is different to doughiness – a sticky dog may be free moving when it moves, but then sticks due to too much eye, probably combined with weakness. A doughy dog may not be sticky, but will always move in a doughy fashion.

Also be careful not to be fooled into thinking that a dog is steady when its steadiness is simply due to poor conformation and poor movement, or excessive eye.

Minimal activity

'Minimal activity' is the dog's ability not to do anything more than needs doing, not taking one superfluous step. *It is the opposite of 'busyness'.* Most dogs do far too much moving about.

The good dog doesn't make work for itself – as soon as it sees the sheep are under control it stops moving (however, this shouldn't come from excessive eye), and only starts moving again if it needs to. Most dogs get the sheep under control, but can't remain still, so they move and hence make more work for themselves. They do the same thing when balancing sheep – instead of holding the exact point of balance they are itching to do something, so they move off the point of balance and create more work.

Be careful not to confuse minimal activity with steadiness. You can have a steady dog that lacks minimal activity.

Without this minimal activity the dog cannot have real 'distance' or 'balance' (discussed in Chapter 4).

Controlled force

'Force' is the ability of a dog to move stock. The majority of three-sheep trial bred collies lack force (they can't move stock), and thus many are fairly useless at practical work.

On the other hand many Kelpies force in the wrong place and in a fizzy, erratic manner, and many of them are always pushing into stock and lack distance and balance, and so you cannot allow them to work on their own. (This is despite what is sometimes claimed for them being 'do-it-yourself kits' and not 'push button dogs'). You have to be continually stopping them or calling them off stock to let the stock draw along quietly, and you hesitate sending them to shepherd a wing or straighten the mob because they will rush around close and create more problems than they fix. This is one reason why a lot of farmers carry their dogs on the bike and work the stock themselves, and only let the dog off when absolutely necessary. However, the best Kelpies are fairly good.

At its most basic the forcing instinct is seen in the dog that, when confronted by unmoving stock, takes that as its cue to move in towards them to get them moving again.

Many dogs, when stock turn and face them, begin showing more eye. Then, when the stock move off, they follow. Instead, the dog should force when the sheep stop and face it, and back off when they move.

Strength is different to force, although it is an important aspect of it. Strength is simply a lack of fear. The strong dog is not afraid of stock, even aggressive stock such as cows with calves at foot. Real force is a combination of the desire to force, and strength.

There are plenty of weak dogs that may have lots of force on easy stock. They may rush in and bite or bark, and fizz and bounce around, and constantly work close, and so on. But when the going gets tough and stock really test them out, they have nothing.

A few strong dogs

Vin Stapleton's good dog Will's Cobber (Castleton Herbie × Will's Maggie) was one example of a strong dog – I have seen him repeatedly charged and hit by a sheep, and his only response was to turn his head sideways to take the blows on his shoulder. The sheep did this a few times before turning and walking off.

I have heard many stories about very strong dogs. A number of them relate to Jack Hiscock giving demonstrations with a dog working a single pig, one with Miller's Mighty (Yulong Russ × Moorlands Tracey) forcing a pig backwards up the ramp of the shearing shed, with the door into the shearing shed closed, and another of Cavanagh's Esjay (Cavanagh's Craig × Cavanagh's Teena) working a pig in a stable. I have heard from a number of people that Esjay 'was a magnificent forcing dog'. What they mean is that he was very strong with great controlled force – the ability to force and hold at the same time, steadily.

Will's Cobber (Castleton Herbie × Will's Maggie)

Sid Cavanagh's dog, Barravore Jim, was also exceptional in his ability to cover and force simultaneously.

However, you can have dogs that are very strong, but may lack force (the ability to move stock), perhaps because they have too much eye, or are too doughy in their movements, or simply have no inclination to get in and shift stationary stock.

The best force

The best force is the combination of real strength (lack of fear) combined with the

Sid Cavanagh with Barravore Jim (Miller's Pete 2nd × Timaru Belle)

desire to force, *but this force must be controlled.* How often do you see people penning up in shearing sheds or working sheep in the yards, with dogs with plenty of force, but the sheep are piled up in the opposite corner to that required? Or out in the paddock with the stock going every way except the right way?

A good analogy is to think about when you are splitting tough firewood. If you choose the right place to hit with the axe, and are accurate enough to hit the same place each time, you will split it in the easiest manner. But if you choose the wrong spot, and your aim is all over the place, you can use a heap of effort and not get very far at all (and make a mess of the log) – many forceful dogs are like that.

Anyone can breed forceful dogs; the trick is breeding dogs that force CALMLY in the RIGHT PLACE, with great holding ability and distance.

For most mob work in the paddock, force simply involves the strong, steady but free moving dog moving closer to the stock (usually without any eye at all), which then move off. However, at times more is needed, such as when putting sheep over a creek. It is here that the importance of 'holding' while forcing should become obvious. The strong, good covering dog will force sheep better than anything, and if it also throws in a well-timed bark or two so much the better. This sort of continuous, tight-holding force is ideal, but very rare.

All the force in the world is worthless if the dog doesn't hold while it forces, because the sheep keep gaining ground on it by breaking back past it, and any ground that the dog had gained is lost. The force is diluted or dispersed.

Dogs that rush in, or around, to force, then break back out and go back into holding mode to retrieve the breakaways, are not ideal. This is associated with an excitable temperament and too much eye, and a lack of real strength and steady force. Such dogs either hold, or they force, whereas the two should be one combined action.

The best dogs hold while they force, due to exceptionally strong, direct, walk-in force with minimal eye, combined with great 'cover'. They force directly rather than sliding around. Some 'bounce' comes from the dog getting in close to stock when forcing soundly in this manner, and then covering fast due to exceptional 'short cover'.

A good dog also understands to come back behind the shoulder of a sheep to send it forwards. Moorlands Georgie was one dog with exceptional short cover, and she would make a little move just back behind the shoulder of sheep – or walk in behind the shoulder – to send them forwards, then move forwards to control and turn this forward momentum. *Good short cover will move sheep very effectively.*

Many dogs slide around towards the head as they force, such as when drawing a mob in towards the handler. The dog comes around and forces on the wing, which generally causes the wing to race along and ring around the lead, which turns the mob and stops it. The dog then races back around to the other side and does the same thing again. Also the sheep or cattle often break back behind the dog.

In contrast the good dog drives the mob from behind and forces on the point of balance or on the tail, and can move large numbers of stock without doing much at all, and move them steadily. *An ounce of force in the right place is worth a pound of force in the wrong place.*

Of course there are times when a dog needs to run up and force on the wings, but this should only happen if the handler commands it to, or it has learnt through experience that this is required at times. *In 95 per cent of cases the dog should force on the tail or on the point of balance*, not rush up the wing to force.

Force should only be used when needed, and it must be balanced by distance, and by break, and by steadiness. The good dog forces only when needed, then backs off and works steadily from a sensible distance.

Bark

Bark is useful at times, however, it should be used like eye – only when it is really helpful.

Mostly barking is greatly overrated. It often causes more harm than good when used by people who always try to rush stock, but only take longer as a result. *Make haste slowly.* Often by keeping quiet, and allowing stock room to flow, everything happens much more quickly even though at first glance there may seem to be less happening.

Good stock work is devoid of 'fuss'.

However, barking is useful in many situations, such as with a dog on the tail of a mob on a road when most of the mob is unaware of the dog's presence, or at odd times in the yards, or forcing stock over a creek. It is preferable to the dog biting.

Bark is also useful when working cattle, if used in moderation at the right moments. For example, when blocking or turning a mob of running cattle (and there can be a lot of momentum in such a mob) the lead or shoulder animals may see the dog but those behind them can be unaware of its presence; a few well-timed barks can alert the entire mob to the dog's position and so more effectively stop or turn it, as can a couple of well-timed cracks of the whip.

The problem with most barking dogs is that the bark is a result of (or associated with) an excitable or ratty temperament, or weakness. Give me a calm, silent, strong dog over a weak, excitable, barking dog any time, but the calm, strong dog is better if it will bark when required.

So, if you are trying to breed some bark into your dogs, always try to breed from calm, strong dogs that bark sensibly with purpose, *as they force*, not the type that stands back and barks, or yaps aimlessly or out of weakness.

Bite

As far as bite is concerned a dog should only ever snap at the nose (if sheep or cattle attack) or the heel of sheep or cattle, and should never hang on. It should do so calmly and fearlessly, and not rush in due to weakness.

A dog that bites, and particularly one that hangs on, cannot hold while it forces, and so will never be a good forcing dog.

Force summary

The way a dog forces is one of the most important aspects of a high-calibre, practical working dog. No matter how good a dog may be in other areas, if it lacks force it is of very limited value.

Real force comes from a combination of outright strength and the desire to force, which can include some purposeful bark and controlled bite. But for the most part it is simply a steady, strong, good covering, free moving dog moving closer to stock *without hesitation*, and meeting any challenges. 'Mob cover', 'short cover', 'drive' and 'balance' are vital components of high quality, exceptional forcing ability, so that the dog holds while it forces, and forces in the right place.

A dog with good force and with distance will only force when required, and will then back off and take stock quietly. A dog that is always putting pressure on stock is nothing but a nuisance.

Temperament summary

Temperament is at the heart of every sheep dog's ability. It determines whether the dog has minimal activity and steadiness, or whether it is busy and excitable. It determines whether it is strong, or weak. It determines whether it is calm and will think about what it is doing or just rush into things. It affects whether it uses too much eye when that is not needed.

Also it determines how easy the dog will be to control. A calm, steady dog with minimal activity will be easier to control, even though it should be hard and strong. If you want to breed dogs that are controllable you should breed CALM dogs with great heart, not soft or half-hearted ones.

A dog's tail is a good indicator of its temperament. If its tail is flying around then it is fizzy. If its tail comes up when it is walking in or holding then it is tense or weak. *The dog's tail should hang down relaxed at all times.*

Temperament and work are inseparable. You cannot have a good dog without a good temperament.

3

The holding traits

'Holding ability' is the ability to control stock. A good holding dog is often described as having a 'grip' on its sheep. However, some dogs may get a good grip on their sheep *when the sheep stay together*, but have no idea of working them when they split up. Thus the other aspect of holding ability is that the dog must have the instinct to hold the stock together, *and very few do.*

A dog needs exceptional holding ability (amongst other things) to master this class of country

There are two main, distinguishable types of holding ability. The first is based on weakness and tension. Many weak or tense dogs (sometimes described as being 'nervy') appear to have a good deal of holding ability in some situations. However, this is mostly due to the weakness causing them to keep out off the stock, and weakness or tension causing them to show a lot of the wrong sort of eye and caution.

In contrast, the ideal holding ability is seen in the calm, strong dog that has a great desire to keep all the sheep together and under control. The traits that I consider to make up this holding ability are:

- heading instinct
- freedom of movement
- width
- hold together
- breakaway break
- square break
- short cover (squarely)
- mob cover
- eye
- the natural stop.

Heading instinct

The most basic holding instinct is the instinct to head moving sheep. However, the dog must head the lead sheep of a strung-out mob, and not cut in part way as many do.

This basic 'heading instinct' is different to a dog 'getting to the head', which means that the dog gets right around in front of sheep without stopping short of the head; that is discussed next.

Freedom of movement, and 'getting to the head (and tail)'

The good dog must move around sheep the full 360 degrees with complete freedom. Also it must do so in both directions – you should struggle to discern any preference in a pup for one direction or the other.

Yet many young dogs only go so far around, and then 'chop in' or hesitate (often their tail goes up in the air as they do so). Even many pure heading dogs will do this. So, although they are obsessed with the head, they don't get fully around in front of it.

You see this trait, or lack of it, in the early training of a pup to 'go back', and when it is learning the directional commands. Some pups go freely and easily around sheep, while others don't – they have a 'point of resistance' as they circle around.

It is this 'freedom of movement', combined with heading instinct and with 'width' or 'break', that gives us a dog that 'gets to the head' (or as I prefer to explain it: 'gets right around in front of the head'). This freedom of movement also results in a dog that will 'get right around behind the tail', when it is combined with 'drive' and 'hold together'. This is just as important if you don't want the dog leaving the tail of a mob behind.

The dog should keep its distance off the sheep as it moves around them (width) and only come in towards them when it reaches the 'point of balance'.

The good pup will arc right around in front of a breakaway sheep when heading it, to turn it back to the mob. In contrast, most pups tend to go only just far enough to stop the sheep, and then either head it again as it tries to get back to the mob, or race it back. The good pup turns a breakaway sheep back, rather than just stopping it from going any further.

This also influences the cast, because the dog that goes right around in front of sheep to turn them back will go right around behind sheep at the top of the cast, and is less likely to stop short of the point of balance (depending on other traits also).

I have a young dog here now (eight months old) Campaspe Bounce (Moorlands Tomie × Cullen's Kiara) high in this ability. He has a good natural cast, and casts wide and deep right around behind sheep with very little training. But another area where you really see the advantages of this freedom of movement and getting right around in front of the head, combined with drive and hold together, is when he is putting a dozen or so sheep over a small creek, with rushes growing on either side of it. The sheep try to break in either direction along the creek in the rushes, and he goes right around in front of them and drives them back to the others and to the crossing.

In contrast, dogs that don't get right around in front of the head tend not to go far enough to drive spreading sheep back, and so end up with the sheep spread out until they are too far apart to control. Be careful though, not to confuse this freedom of movement with a dog that simply overruns.

So look for the pup that breaks wide off a breakaway sheep to get in front of it to turn it back to the others, and which has no point of resistance or chop in as you move it around sheep. This getting to the head is very important, and is one of the first things I look for in a pup.

Campaspe Bounce (Moorlands Tomie × Cullens Kiara). Bounce has 19 crosses of Yulong Russ

Width

I define width in a dog as the tendency to give sheep some clearance as it goes around them, all the way around them. Some dogs tend to go very close to stock when going around them, or even go through a mob, while others keep some clearance as they work.

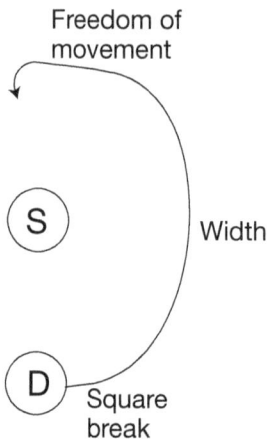

Figure 3.1 Square break with width and freedom of movement

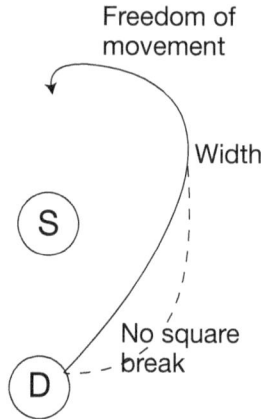

Figure 3.2 Width but lacking square break

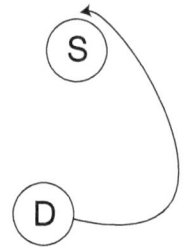

Figure 3.3 Square break with poor width

Width is different to 'breakaway break' (discussed later), because breakaway break tends to be shown when sheep run or are wild (or all the time if the dog has too much eye). Also it is different to 'pressure break' because that tends to be shown only in conjunction with eye and weakness when working on the head of sheep. Width is shown even on stationary sheep whether or not the dog is using eye, *and around the tail as well as around the shoulder and the head*.

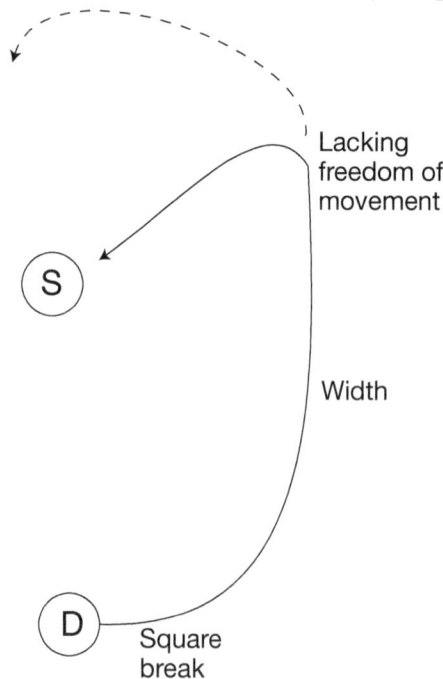

So a dog with good width is simply a dog that curves around its sheep giving them good clearance for all 360 degrees.

'Square break' helps in getting the dog started right on its width, as shown in Figures 3.1–3.4. Freedom of movement is also important because otherwise the dog will cut in before it gets to the point of balance.

Figure 3.4 Square break and width but lacking freedom of movement

Hold together

Next comes a trait that is *very rare*, and yet essential for high quality work. I define 'hold together' as the desire of a dog to keep sheep together, and to try to put them all back together if they split up. From a young age and without training, the dog should keep a good distance out around all the sheep, even if they separate, *and should work the sheep as a mob*.

This ability is different to break or width. A wide-breaking dog can appear to always keep out around all the sheep because its break carries it wide around them, but the reasons for it doing so are different. The pup with hold together doesn't necessarily work wide; it might work right in close if it lacks width or break, so close that it causes sheep to split off. Yet it always retrieves those sheep and *has the desire to keep them all together*, and not to leave any behind.

This is one trait that particularly needs to be tested on sheep that will spread apart, although you can get some idea by just watching what the pup is looking at – a good pup will be watching all of the sheep and keeping them all in its mind, whereas a poorer pup will at times focus on only a couple to the neglect of the rest (particularly if they separate from the others to some extent), or will focus on the lead to the exclusion of the tail.

Remember not to confuse break – or a dog working wide out of weakness – for hold together.

Break

Break in all its forms is basically a dog that breaks out wide as it begins to move around sheep in one form or another and in one situation or another, instead of going straighter towards and around them.

This is perhaps one term that is more confusing than any other, because it has various forms depending on the bloodline. There is an infinite variety in the way dogs move around sheep, and the various situations in which they may, or may not, use one form or another of break.

So from a breeding point of view it is no good saying that you want to breed dogs with break, unless you define exactly what sort of break you want.

Pressure break

The most common sort of break in three-sheep collies is what I call 'pressure break'. It occurs in weak or tense dogs that have inherited this type of break (not all weak dogs have it) in combination with 'pressure eye'.

The dog breaks out and arcs wide on the head of sheep when blocking them. However, it only does so when it is using eye, and the only time it uses eye is when it is working on the head of the sheep under pressure. Such dogs often shoulder a breakaway, because they don't feel pressure when sheep are running away from them.

This pressure break can give a semblance of casting ability, because the dog tends to break out around sheep, but it does so through weakness. It is undesirable.

If you attempt to breed a strong dog with break, by breeding the weak, pressure-breaking dog to a strong dog lacking in break, you will be disappointed – you will get weak, wide-breaking pups and stronger shouldering pups, because such break is due to inherent weakness and tension.

Breakaway break

Another type of break, desirable this time, I refer to as breakaway break. This is where the dog gets wider when heading a running sheep, instead of cutting in close and

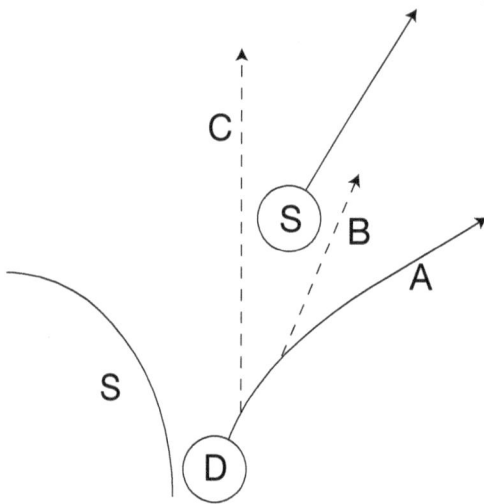

Figure 3.5 (A) Breakaway break (B) Shouldering (C) Cross-heading

shouldering it, or even cutting in between it and the other sheep (which many dogs do, even some wide-breaking dogs – a dog should never cut between one sheep and the others, but always keep out on its own side of all of them). See Figure 3.5.

The slower the sheep are going, the less the dog breaks off them (if it has the right sort of eye), but when working wild sheep it will break very wide in order to keep them under control.

This is one type of break that I look for. When I try pups on sheep and give them their head to do as they please in a small paddock, I look for a pup that will break off a breakaway sheep and get in front of it to turn it back. *If a dog lacks this break it will never handle wild sheep.*

A dog showing eye tends to use more of this breakaway break (assuming it has inherited this trait) because it is imagining (due to its tension and excitement) that the sheep are going to break. This is fine when working touchy sheep, but if a dog has too much eye then it may show a lot of breakaway break even though the sheep are dead quiet. Such dogs won't work with minimal eye in close to sheep, or force calmly. They use eye and break even when working quiet or doughy stock, then rush to force (if they force at all).

Drawbacks of dogs with too much break

Some dogs with pronounced break can break with too little provocation, such as when holding sheep against a fence. They break off their sheep as soon as the sheep make the slightest move, and hence they are no use at any forcing/holding work.

A good dog should show strong 'block and hold' when holding and forcing on the head, and only break off if sheep get past it, to retrieve them cleanly. (Block and hold is basically a high level of 'short cover' combined with strength. Also the dog should cover its sheep squarely, because if it moves directly towards them some will get back past it.)

Also some excessively wide-breaking dogs can keep giving ground – they break off a breakaway sheep but continue out from it, and can end up simply running along out wide, parallel to the sheep but slightly ahead of it. Or, if they do get in the sheep's path, they can end up running away from the sheep (leading it) due to the excessive break. Such dogs are the type that make 'lead dogs', but they aren't much use otherwise.

Some dogs that break excessively wide do so not because they have too much break, but because they lack the *opposing* traits to limit it, such as strength, or force, or block and hold. When breeding working dogs, it pays to keep this in mind – you generally shouldn't try to breed dogs with less break, but rather with more strength and force and block and hold.

Square break

'Square break' is the way that a dog turns at right angles before moving around sheep, instead of moving more or less straight towards and around them (see Figure 3.6). A good way to describe it is to think of a dog working a big mob of sheep. When the dog begins to move around them it must turn squarely simply due to the shape of the mob. But when the dog is only moving around a few sheep the shape of the mob doesn't force it to break square like this, *but its breeding should.*

Figure 3.6 (A) Good square break; (B) lacking square break

So, regardless of the number of sheep being worked, the dog should break square as it begins moving around them. It should do this even when showing no eye at all, and whether the sheep are stationary or not. *It should also break square even when the sheep are moving away from it.*

Square break is the start of width, but a dog with square break can initially break square but then cut in towards the sheep, because it doesn't have width (see Figure 3.3).

Also a dog can have this square break but no breakaway break, so that it shoulders a breakaway sheep.

Break summary

So you can see that break is a fairly loose term with many variations (including others I haven't mentioned). What is wanted is a dog that breaks squarely as it begins moving around sheep, and that breaks wide off breakaway sheep. It should break wider the wilder the sheep are, and less the quieter they are.

A dog should always break *outside* of a breakaway sheep – many strains of dogs will cut in between a single sheep and the mob at every opportunity, and cross-head them, instead of always breaking outside them. This is a bad fault.

Good dogs should have both square break and breakaway break.

Cover

'Cover' is the *way* a dog moves *around* sheep in order to control them and to keep them together. This is in contrast to moving in directly to force the sheep. Some dogs are described as having 'all cover and no push', which means that they are always moving around sheep and never moving in directly to shift them. Very few people understand what good cover is. In my opinion, many dogs described as having good cover actually have very little.

I have separated cover into two main divisions, 'short cover' and 'mob cover', which can be further broken down.

Short cover

Short cover is basically a dog's ability to read sheep well, and to react instantly to their *slightest* movement with a corresponding precision movement of its own. Mob cover

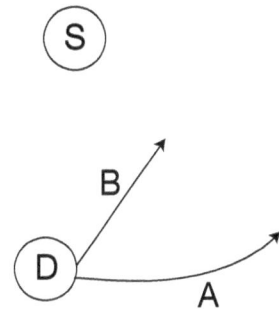

(described below) is the dog's ability to cover a mob, while short cover is the ability to cover the finer movements of individual sheep within the mob (although short cover is an important component of mob cover also).

The better short cover a dog has, the earlier it will *calmly* react to the sheep's *slightest* movements, and it will do so with more *precision*. Because it reacts earlier it doesn't need to move so far, perhaps only a few centimetres (an inch or two), or even just a sideways movement of its head. Because it reacts with precision, it doesn't then overrun. Most dogs respond late to the sheep's movement, and therefore have to move further to regain control.

The good dog should show exceptional short cover even when calm and using very minimal, or no, eye. This contrasts with some dogs that only show a lot of short cover when they are keyed up and showing a lot of eye.

A high degree of short cover gives a dog 'footwork' when working touchy sheep, but some dogs have a lot of so-called footwork constantly, even when working quiet sheep. This is undesirable because that sort of footwork comes from a busy or excitable temperament – it is a bit like a dog with 'ants in its pants' – it can't stand still. The good dog has very precise footwork, but doesn't dance around on its toes for no reason.

The best short-covering dog is the one that will cover its sheep from the point of balance, even when it is relaxed and using minimal eye. Short cover is just as important for high quality mob work as for working small numbers.

Square cover

By 'square cover' I mean square *short* cover. Some dogs with square break may cover a *mob* squarely, and yet not do so when short-covering sheep within that mob – that is, when covering individual sheep they move straight in and get closer to their sheep ('encroaching' as I once heard it called) – they are drawn to their sheep.

Such dogs will cover (short cover) their sheep straight, and lose the mob's point of balance, and then break back out to find the point of balance again. They cover straight, then break.

Instead, the two should be one combined action that I call square cover. The dog will move in towards the sheep to move them, perhaps only one step, then step sideways to cover them. What is required is a dog that covers its sheep squarely while it maintains its distance off them, even when covering on the tail.

Real square cover is affected by many things, including 'distance', 'minimal activity', 'square break', 'short cover', the 'natural stop', 'hold together', and the right type of 'eye'. It is very rare.

Mob cover

Mob cover is the ability to cover a mob and hold it together, particularly when sheep are wild and trying to split in every direction. It combines various other abilities, such as heading instinct, width, hold together, short cover, breakaway break, and square break.

You rarely see a dog with good mob cover, because most lack hold together and short cover, but there are a very few about. Most people have never seen a really good covering dog.

Freedom of movement and the ability to get in front of the head is also important to keep wild, splitting sheep together. The dog that doesn't get far enough around in front of sheep as it heads them tends to allow the mob to spread apart, until it can no longer cover enough ground fast enough to control them all. The same applies with the tail – if the dog doesn't get right around behind the tail to move them up towards the mob, the tail will get left behind.

Cover summary

Cover is vitally important in almost every aspect of a dog's work, including forcing, and when working in yards and shed. The good covering dog doesn't allow breaks to happen, and gets the job done much more quickly and easily. *Short cover is very effective in moving stock*, as well as in holding them.

However, very few dogs have really good cover, and if they do they mostly won't cover and force at the same time. The good dog must have a high degree of cover combined with strong force, a minimal amount of the right type of eye, and a calm temperament.

I have not seen another line of dogs come within cooee of having the outright cover of the best Yulong Russ blood collies. Years ago Sid Cavanagh informed me in a letter that most of his dogs were from Yulong Russ bloodlines, and that he thought it was the best bloodline. I have found him to be right ever since. There are a couple of lines of Kelpies left (all somewhat related), which I consider have the potential to produce high quality dogs, but none I have seen can match the Russ blood dogs for outright cover.

Eye

There have been many ridiculous definitions of 'eye' over the years, the common one being that it is the power of a dog to control (or move) sheep, simply by looking at them. In fact, eye is simply the outward manifestation of concentration and caution. In the wild it is shown by the big cats, or wolves, stalking prey; that is, trying to creep up unobserved, or trying to anticipate what the animal is going to do.

Dogs show eye as they get more intent and focused and cautious, or frightened.

The amount and type of eye a dog uses is vital in its effectiveness as a working dog. Too much, or the wrong type, and the dog will lack force and cover, be difficult to move around, and will cause sheep to stand and look at it. The dog will also rush to force (if it forces at all). On the other hand, too little eye when it *is* needed and the dog will be loose in its work, and have trouble handling touchy situations.

A three-and-a-half month-old Kelpie pup showing eye

Pressure eye

I believe there are basically two forms of eye, or two motivations behind the eye. The first I call 'pressure eye', in that the dog only shows eye when under direct pressure. This is due to weakness and fear. It mostly occurs when the sheep are facing the dog.

Such dogs generally don't show any eye when working on the tail, and often show a fair bit of pressure break around the head, showing strong eye, but will cut in across the tail showing no eye at all. Such dogs also tend to show a fair amount of 'stickiness' and can be hard to move around.

This type of eye doesn't breed on if you attempt to remedy the weakness. If you breed the weak, pressure-eyed dog to a stronger dog lacking in eye, in an attempt to get strong dogs with plenty of eye, you will be disappointed. You will end up with weak pups with eye and stronger pups without it.

People breeding for eye and style are often unknowingly selecting for weakness and/or tension, because the most common sort of eye in this country (not in the UK) is associated with these traits. If you want to breed strong dogs with eye and style, then generally you must breed from strong dogs with eye and style.

Some people with dogs that show eye and hold, and then lose the eye as they rush to bark and force, think that their dogs have the 'right' type of eye. But in fact they do not. The reason they rush to force is because they have too much eye, too much of the time. They are the same in principle as the strong-eyed three-sheep trial dog that won't walk in, and then dives in out of weakness. If they had the right sort of eye, and force and strength, they wouldn't rush in the first place.

Dogs with pressure eye generally won't use it on chickens, or a cat, or a rabbit, etc., because it's only evident when they are frightened or tense to some extent.

Free eye

The other type of eye I call 'free eye'. This is what is required (in strict moderation). It denotes good concentration but is not brought on by fear. This type of eye more often breeds on.

The dog will show this sort of eye even on the tail of sheep when they are moving away from it, or anywhere else, or on the chooks, or on the cat. Most UK dogs have this sort of eye, and some Kelpies do, whereas most modern three-sheep dogs have pressure eye.

A dog should use minimal eye most of the time when working a mob, or quiet sheep, or in the yards, but show enough concentration when working touchy stock. *This is vital in the practical working dog.*

The calm dog will continue to breathe even in tight situations when showing a fair bit of eye, and carry its tail hanging down relaxed, whereas the tense, high-strung dog will clamp its mouth tight shut, and possibly raise its tail. A dog that raises its head when showing eye is also generally weak or tense (like the dog that raises its tail).

The following quote from the late C. Brewster Macpherson of Scotland (breeder of the famous Border Collie Macpherson's Moss, who was imported to Australia), written in the 1920s, gives a good sense of the right type of eye:

> A 'strong eye' is essentially a quality of the border dog – and, so far as working a few sheep in a trial goes, it is useful and attractive. In practice, however, in hill work I have never been able to discover its advantages.
>
> There, by 'eye' I mean something very different. A good eye for me means the wonderful eyesight possessed by the best hill dog – the bold sweeping

glance that sees everything, that is always forward to the end of the flock, and misses no straggler in the far distance; the "strong eye" of the trial ground need show none of these things, and how often have I cursed that powerful optic when it has fixed its baleful vision on, and glued its possessor to, the first pickle of sheep it falls on in a distant cast, to the exclusion of the wildly scattered flock beyond.

The good dog should look fairly 'plain' in most practical situations, only tightening up when required. It should have the ability to work stock to a high standard with no eye at all, with *a touch of eye just being the icing on the cake*.

The good dog works like a person – calm, plain, and thinking about what it is doing, and only tightens up or speeds up when it really has to. For most practical work I would prefer a dog with too little eye to one with too much.

The natural stop

The good dog should arc around sheep and stop itself on the point of balance, before summing up the situation and deciding either to walk in to force, or to move again to cover, or to remain in that place. However, many dogs will go around to the point of balance (if they have good width and freedom of movement) and then come straight in without pausing first.

Some people refer to this 'natural stop' as 'balance', but this is wrong. Although the dog may stop on the point of balance as it goes around sheep, it might not *maintain that point of balance when the sheep are moving* (balance is the ability to maintain a *moving* point of balance) – it may over-head or lack the short cover or minimal activity to maintain the moving point of balance, and therefore will lack balance. Also a dog can possess the natural stop and not stop on the point of balance, but only stop on the head. So the natural stop is different to balance, although it is one aspect of it.

The natural stop also helps make a dog easy to control, and is one aspect of real distance. This is what gives the natural lift, but it must not come from weakness or excessive eye.

Holding ability summary

In summary, we require a dog that holds sheep together and keeps a good grip on them, and when this is combined with driving ability (as explained in the next chapter), we have the ideal combination.

This means that the dog with heading instinct should also have width and hold together, so that it doesn't split sheep and if they do split will try to put them back together again. The dog should also have breakaway break to break off sheep that run – it should get wider as it goes, and break out to get around in front of the sheep, to 'get right around in front of the head' and turn them back.

It should move completely freely all the way around the mob, without chopping in or getting impatient or frustrated, and pause naturally on the point of balance before deciding what to do next.

It should also have a great ability to cover its sheep, reacting quickly but *calmly* to their slightest movements with a precise movement of its own, to keep them together and under control. A dog that takes its eyes off sheep in pressure situations will not cover them well. The dog should cover squarely, and not encroach on stock as it covers. It should not break wide when it is forcing on the head, but should show strong block and hold.

Finally, it should use enough eye when sheep are wild and touchy, and yet be free from any stickiness – free eye, and should show none or very little on a mob (depending on the situation) or on quiet stock.

The subdivision of traits

Some readers may wonder why it is necessary to break the dog's ability and traits down into so many areas. The basic answer is that I wanted to know exactly what makes a good dog, so that I could select for the right things.

For example, it was no good simply saying that I wanted a good holding dog, because a dog can hold stock for a number of different reasons. Only if it did it for the right reasons would it be a top holding dog, and have potential to be a good dog in other areas also.

Many breeders fight against themselves because they don't understand this fact, or the traits, properly. They try breeding a weak holding dog to a strong forcing dog (with less holding ability), in an attempt to breed a good holding dog with good force. But if the 'holding ability' is based on weakness, as it often is, then what they are trying to achieve isn't possible.

Then there are other breeders that don't bother, but simply breed dogs lacking holding ability because they think that real holding ability cannot co-exist with strong force.

Therefore I had to break holding ability down further, so that I knew which traits to look for to give real holding ability that wouldn't preclude good force (or drive).

To give one example of why I then broke holding ability down into the traits I did, consider width. I thought that this was a good trait to have in a dog, but I wondered whether it was due to weakness. If width were due to weakness then its presence would exclude real strength or force. But then I had a number of dogs through my hands that had plenty of width, and yet were also strong with good force. So I realised that the two were not mutually exclusive.

And I had dogs that kept out around all the sheep in easy situations because they broke wide, but I came to realise that such break was due either to weakness, or to excessive eye, or simply because the dog constantly broke too wide. So those forms of break were undesirable, because they precluded the dog from having good force, or from having the ability to 'block and hold' tightly, whereas width didn't.

But I found that you could have dogs with width that would still leave sheep behind (as can dogs with break). So I realised that simply because a dog kept out around stock didn't mean it really had the idea of bringing all of them.

I also had some dogs that had this desire to bring all the sheep (hold together), and yet lacked width, and so I realised that the two were separate traits. But even though these dogs with hold together wanted to bring all the sheep, the lack of width meant that they would sometimes cause them to split up in the first place. So both traits were necessary.

Another example is casting ability, discussed in Chapter 5. A dog can cast to some extent for a number of different reasons, but only if it does it for the right reasons will it be a *top* casting dog, and will it be possible for it to be a top dog in every other area also.

So this process was gone through with all of the various traits, over many years, tested against the many dogs I owned, and all those I saw work, in all manner of situations, until I came to a consistent and comprehensive picture of exactly what is needed to make the perfect dog, in traits that are not mutually exclusive.

Thus, I have tried to break the working dog's ability down into its most basic, and more or less stand-alone, units, all essential in order to breed the perfect dog.

4

The driving traits

Driving is the opposite of heading, in that the dog works the sheep from *behind* them, instead of from *in front* of them. Most dog breeders in this country fail to consider driving ability ('drive') when breeding dogs, and don't understand the vital role it plays.

First of all, the majority of stock work involves driving stock. For example, droving stock along in the paddock is mostly driving work, rather than heading. The 'draw' also should involve the dog driving the stock to the handler (working them from behind), whether with one sheep or a thousand.

The only difference between drawing stock to the handler, and taking them away from him, is the handler's position. To the dog there should be no difference at all – both involve driving, or balancing, the stock in a certain direction, as opposed to heading them.

Consider the following old quote from the late Mr Herb Morrice about a young Kelpie he had:

> When close to her sheep she would not bite and she had a good idea of holding together and driving. (Kelley 1942)

The dog must have both the natural ability to hold stock, and to 'work them from behind' or drive them. Breeders must look to the dog's driving ability if they want to breed top dogs, and this is a big shortcoming in most Australian dogs.

Also, I have included 'distance' and 'balance' and 'brainy stock work' in this chapter on driving ability, because drive is the foundation of all three.

One reason people have trouble breeding dogs with both a high level of force and a high level of distance, is because they don't understand the role drive plays. They find that they can either breed a weak dog (or one with too much 'eye') that works wide and cautiously, or a strong dog that doesn't.

The answer lies in understanding drive, and distance, and the right type of eye, and the problems associated with over-heading and fizz, as explained in this chapter. To get a very strong, forceful dog with good distance, it must be calm and 'work its sheep from behind'.

Nor do most people understand distance. They think that it is just a dog working wide, but this is not so. A wide-working dog can still be drawn by sheep, and chase them along at a full gallop when they are moving away from it. This sort of wide work is mostly a result of weakness. Real distance is based on drive (and other factors) as discussed below.

Drive

So what is 'driving ability', or drive? In its pure form it is seen in the dog that simply follows behind sheep, making no attempt to head them or to stop them, but simply follows them wherever they go. Many cattle dogs (heelers) were of this type, but even they were valued if they would also head breakaway cattle.

Sometimes dogs with drive have been described as 'working their sheep from behind', as I have heard said of the great Yulong Russ. In order to be a good dog a dog must possess this ability, combined of course with exceptional holding ability.

The dog with natural drive (and distance) will not attempt to ring sheep on the draw (which is a bad fault), even when young and with only minimal training. Also it will be easy to teach to cast into a tight corner and to go behind sheep and draw them out, without simply holding them into the corner like the 'mad keen heading dog' will, even though it may possess exceptional cover and holding ability. And when droving sheep along in the paddock, you don't have to be continually calling the driving dog back, as you do the heading dog as it runs too far up the wing all the time.

Pure heading dogs also dislike forcing on the tail. They might slide around and force on the head or on the wing, but not on the tail. It is as if you are trying to press the same poles of two magnets together, which repel each other. If you try to force the dog into the tail, to force there, it slides around towards the head.

It may seem strange to those who are used to pure heading dogs, but outstanding heading ability, cover and holding ability are perfectly compatible with this driving ability. In fact, to have a top dog, *you must have this combination.*

The driving dog, if it is calm, can have the idea of taking stock somewhere, and so may have purpose. *The over-heading dog can never have purpose.*

Some UK Border Collies are obvious examples of driving dogs with good heading instinct (although most lack balance and mob cover). They don't overrun like a pure heading dog does, which is so evident when working one sheep. A driving dog (like those from the UK) will handle one sheep out in the open without overrunning, whereas a pure heading dog overruns too far towards the head continually, and the sheep gets back behind it, or baulks.

The two types of dogs tend to work from two different bases. The heading dogs base themselves on the head, and that is where they will stay unless commanded or trained off it, to allow stock to move. In contrast the good driving dog bases itself on the tail, and *moves itself* from there towards the head as required. This makes it a more *natural, thinking worker* than the pure heading dog.

Distance

Distance can be defined as a dog that isn't always boring into sheep, and that isn't drawn by the sheep. It is different to a dog that simply works wide, mostly out of weakness or excessive 'break'.

Sometimes dogs with some elements of distance are described as being good 'shepherding' dogs, in that they will keep off stock and just guide them or hold them. The opposite is the *pushy* dog, or the dog that *bores* into its stock all the time.

The main things that work against a dog having distance are an over-developed heading instinct, excitement, busyness, and a lack of drive. To put this another way, for a dog to have the greatest possible distance it needs:

- drive
- calmness
- 'minimal activity'
- the 'natural stop'.

Boho Misty had the most perfect distance I have seen. She would work right back off, and wide of, wild sheep, and in close to quiet sheep, and adjust the way she worked to suit the situation. (I have a high-quality pup here now, Glendaloch Craig (Cymru Coon × Campaspe Meg), who is showing similar potential.)

Misty was a perfect natural driving dog (but also a perfect heading dog, and I don't say 'perfect' lightly), extremely calm, with great minimal activity, devoid of any busyness, with a perfect natural stop (and lift), and with good 'square cover'.

In contrast, imagine the pure heading dog drawing a few sheep to the handler on its own initiative, that is, being let do its own thing. It has been taught to stay on the far side of the sheep by being repeatedly chased back there each time it attempts to come around the handler's side to head the sheep.

When drawing the sheep to the handler the pure heading dog will tend to weave from one side to the other behind the sheep, looking for the head on each side, and only staying on the far side through training. It will generally move too far towards the head on each side, and so swing the sheep from side to side.

Heading dogs hate getting left behind by sheep moving away from them, because they are losing contact with the head. Therefore when sheep run, the dog runs to keep up. If the sheep stop, the dog stops, which is just about the opposite of what you need.

Instead, the good dog forces when sheep stop, then backs off when they move.

Thus the pure heading dog cannot have distance. Neither can it have balance (discussed below). It is simply not possible. Only the calm dog that works its sheep from behind can ever have distance.

The occasional good dog also has the ability to adjust the way it works to suit the occasion, and this is more than just a case of using no eye on a mob and some eye in touchy situations. Boho Misty was once again an outstanding example, working wide with wild sheep and close with quiet sheep, and adjusting almost instantly to their behaviour.

I have a five-month-old red-and-tan Kelpie pup now, which is showing signs of having fantastic distance even though he is strong and has plenty of 'force'. When first let loose on a few shorn sheep in a half-acre paddock, his first inclination is to come in and force, covering while he does so. But if a sheep manages to get past him he will retrieve it, and when he gets back to the mob his whole demeanour will change – he will now keep back five or six metres off the sheep, and watch them carefully. *He changes his approach to suit the situation.* As he ages, and gets more experience, he will learn to read the sheep's mood more quickly, without needing to experiment first.

These are the type of dogs that make the true all-round dogs – dogs that read the mood of their stock, and change their tactics rapidly. They are very rare.

Balance

I define balance as the ability of a dog to work on the point of balance, in order to move and control sheep in a straight line. It is a trait in its own right, but is influenced by other traits.

For example, it is possible to have good heading dogs with good driving ability that lack balance. They head, or they drive, *but have nothing in between.* The good balancing dog juggles the two roles, particularly if it also has good 'hold together' and 'cover'.

The better minimal activity, square cover, and natural stop the dog has, and also the less one-sided it is, the more perfectly it will maintain the moving point of balance.

So what or where is the point of balance? It is a point somewhere between the tail and the head of a mob of sheep at which the dog needs to be in order to move them in the desired direction. If the sheep are leaning to one side particularly hard, then the point of balance will be further towards the head on that side; however, if the sheep want to move in the desired direction the point of balance will be directly behind them on the tail, which is where the dog should be.

Most of the time a pure heading dog is too far towards the head of the point of balance. I have heard people say of such a dog that 'he has great balance', but this is wrong.

A common sight at three-sheep trials, among some of the more successful handlers, is a pure heading dog working as if it is on an elastic band attached to the head of the sheep. The dog moves to the head and blocks the three sheep, whereupon the handler commands it back towards the tail to give the sheep room to move, then the dog springs back to the head. The handlers use this as a calculated tactic to keep the three sheep moving quietly.

They generally breed a very fast, busy, mad heading dog, with a busy sort of footwork (a bit like a dog running on hot coals – it can't stand still). Because the dog is always moving around very fast it gives the sheep the impression that it is almost in two places at once, and so they give up any thought of breaking away. However, there is no balance involved.

On the other hand a pure driving dog will be *behind* the point of balance.

Real balance requires both heading instinct and driving instinct; without drive you cannot have balance.

To 'drive' and 'heading instinct' you must then add 'short cover', because a dog that is loose and slack, which moves late and then overruns the point of balance, will not balance its sheep properly – they will swing from side to side. Square cover is vital, because dogs that cover too straight lose the point of balance, and then have to break back out to find it again (if they have break). The dog must have some 'free eye' to give it the concentration necessary to read the exact point of balance. Also it must have minimal activity, because any busyness will cause it to move off the point of balance.

So a lack in any of the following areas will reduce the balance of a dog:

- drive
- heading instinct
- square break
- short cover (squarely)
- free eye
- minimal activity
- the natural stop.

The ability to balance a *mob* (or even just more than one sheep, if they split up a bit) also requires hold together. So exceptional balance is a really good indicator of many traits.

Direct precision force

Dogs bred to drive also often have a great deal of direct walk-in force, in particular the UK-bred dogs. Such dogs will force *directly on any point* of the mob.

This direct force is very important for exceptional levels of 'controlled force', and for good stock work in general. The dog should force directly, and not slide around when it forces.

Drive summary – 'brainy stock work'

In summary, the sheep dog must work its sheep from behind, and base its work from there. *Without drive you cannot have balance, nor a good natural draw, nor distance, nor good stock work full stop.* You end up with a dog that runs *around* sheep all the time, instead of forcing or balancing directly where required.

Also consider that without drive you cannot have what is described as 'brains', or 'brainy sheep dog work'. This is something that Rod Cavanagh is a big believer in – he sometimes quotes the late Allan Miller as saying that we should 'breed for brains'.

As I mentioned earlier, Yulong Russ has been described as 'working his sheep from behind'. I have also heard him described as 'the brainiest dog ever to work a dog trial'. You cannot have the latter without the former.

Moorlands Tomie is another dog with very brainy stock work, and he has high levels of drive, calmness, minimal activity, and exceptional intelligence.

Allan Miller working Moorlands Lad (Minnieburn Pedro × Miller's Kim). (Photo courtesy Jean Parks.)

Boho Misty was likewise exceptionally brainy in general work, with exceptional drive, distance, minimal activity, the natural stop, and intelligence. And yet she lacked the real mustering instinct, and this is another example of how separate instinct and intelligence are. Misty was one of the most intelligent dogs I have come across, and with exceptionally brainy stock work, and yet she had no idea of mustering scattered stock (very few dogs do).

With experience, the driving dog can figure problems out (if it is calm with the right type of eye), and drive sheep away from the handler around an obstacle, or down a fence line to find a gate. The over-heading dog will only hold the sheep up against a fence, or against an obstacle.

The pure heading dog mostly cannot work stock intelligently, because working stock intelligently means moving them, whereas *the instinct of the pure heading dog is to always prevent stock from moving.*

Minimal activity, steadiness, the natural stop, and minimal eye are also very important to give the dog real brains, because a fast, busy dog will never be a thinking worker, but drive is the basic foundation.

The heading dog – or not?

Many people refer to sheep dogs as 'heading dogs'. However, I believe this is a very poor choice of words and one that has done a lot of damage to the breeding of dogs, through the false impression it creates in people's minds as to what a sheep dog should be.

It hasn't done any damage in the UK, because their trials demand a dog with good drive, and they tend to talk about 'gathering dogs' rather than heading dogs (even though the vast majority lack real gathering/mustering ability). Nor did the term 'heading dog' do so much damage in the past, because many breeders were also working stock and so this unconsciously tempered the breeding of pure heading dogs. It also hasn't done as much damage to the Kelpie as to the three-sheep show collie, but it has certainly done enough.

The good dog is the dog that works its sheep from behind, from the point of balance or the tail, and yet gets fully to the head when required.

There is a world of difference between this type of animal and the second-rate, over-heading one.

5

Cast and muster

Casting is the ability of a dog to run out to get behind sheep, in order to work them back to the handler, or maybe to then drive them in some other direction.

Macpherson described the big casting dog well:

> It is amid such surroundings that the charm of dog work lies for men – the wide cast where one has to use a glass to pick up the fleeting form among the rocky steeps. (Moore 1929.)

In my view natural casting ability can be broken down into four main areas, which can then be further broken down. These are:

1   the 'natural pull'
2   the 'sighted' short cast
3   the 'blind' searching cast
4   mustering ability.

I have broken casting and mustering ability down as above *because a dog can have one aspect and not the others*. For example, a dog can have a good blind cast and yet have a poor sighted cast, or vice versa, and may or may not have the natural pull, or any mustering ability.

The natural pull

The natural pull is the dog's natural ability to work sheep back to the handler. If a dog possesses the natural pull, it will cast out readily and work the sheep back to the handler with very little training at all.

An oft-repeated cliché is that 'a heading dog's natural instinct is to go to the other side of sheep and work them back to the handler'. Yet how many breeders actually give this any consideration when breeding their dogs, and how many breeders have actually ever owned a 'natural pulling dog'? Not very many, despite their claims otherwise.

If they were to let a young, started dog – eight or so months old – have its head in a 10-acre paddock, it would be just as likely to hold the sheep against the far fence, or chase them all over the paddock, as it would be to bring the sheep back to the handler.

Due to the scarcity of natural pulling dogs, other handlers will say that there is no such thing as a 'natural pulling dog', and that a sheep dog's natural instinct is simply to get in front of sheep (head them) and hold them in one place. Indeed, this is the case with most dogs, however, the natural pull does exist.

This is not to say, however, that a four-month-old pup, when first let out on sheep in a one-acre paddock, should go around behind the sheep and bring them back to you with absolutely no training. If it is any good, it will simply hold the sheep together against a fence somewhere.

It is not until you begin to train the pup, and until it learns to respect you and view you as the boss, that this ability will usually start to show. You break the pup in to 'go back' to stay on the other side of sheep (which is very easy with such dogs), and then you can take it out and send it around sheep 100 or 200 metres away with very little training, and it will work the sheep back to you without a word being said to it. (Campaspe Bounce would cast half-a-kilometre at 10 months of age, wide and deep, and work the stock back to me without a word being said to him.)

And the instinct related to wanting to work sheep back to the handler is a purposeful one – the dog has the right make-up to want to 'do something' with the stock, that is, take them somewhere (to you) rather than just chase them around for the fun of it.

Such dogs are the dogs that will develop real brains in their work, and later on you can divert this purpose from only bringing sheep to you, to doing all kinds of other things with them.

Casting, and the natural pull, in the wild

I have seen a pair of wolves hunting, on film, and it is possible to see where both the natural pull and the casting instincts originally came from. The two wolves spotted a hare grazing away up on a hillside. One wolf crept directly towards it, and then hid behind a tussock some 300 metres this side of it, and waited.

The other wolf cast itself out wide around the back of the hill (probably 300 metres wide of the hare), out of sight of it, and reappeared a couple of hundred metres behind it (you wouldn't see many so-called well-bred sheep dogs match that cast). It then worked its way back towards the hare until the hare took off, and balanced it beautifully in a line for the hiding wolf, which waited its moment and pounced from the tussock.

Heading, pulling, and driving

Yet very few heading dogs nowadays are natural pulling dogs; most heading dogs are just that – all they want to do is get in front of sheep and stop them. So if you send a started pup around sheep 100 metres away, it is likely to simply keep ringing them, and keep them in the one place. You have to teach it to stay on the far side of sheep, to work them back to you and not keep coming around in front of the sheep, by gradually extending the distance you send it, or by sitting it down on the draw each time it looks like ringing the sheep.

This is what I define as a 'pure (or over) heading dog', and this is what three-sheep triallers are mostly looking for – a dog whose whole aim in life is to *stop sheep moving*, to head them. Many Kelpies are the same. With such dogs you have no end of trouble when it comes to driving stock with you, because they always want to go around the lead and block them.

The farmer's common condemnation of such dogs is just – they complain that the dogs won't drive, and I agree wholeheartedly.

To get a natural pulling dog – a dog that naturally goes around sheep and works them back to the handler – it must possess 'drive' (as well as the natural pull itself). In fact this must be the base from which it works. The dog drawing sheep to the handler should be driving or balancing them to the handler, and not simply 'not heading or not ringing them' through training. There is a big difference. The *natural* dog must drive sheep.

The natural pull is a very important aspect of the mustering dog.

The 'sighted' cast

However, just because the dog has the ability to run to the other side of sheep and to then work them back to the handler, does not mean that it will run out *well*, that it will cast out *wide* and *deep* with minimal training. This is where the second point comes in, which is the *shape* of the 'sighted' cast. The pup should go wide and deep right around the sheep to the point of balance, before coming in.

The shape of the cast is affected by other traits, and there are a number of different reasons why a dog may cast 'well' on these sighted casts. Some do so out of weakness and wide 'pressure break', which is obviously undesirable.

On the positive side, 'square break' will get the dog started off right on the cast, and along with 'breakaway break' and 'width' will keep it out wide and deep. 'Scope' will help also. 'Freedom of movement' greatly affects the shape of the cast – it and 'drive' will ensure that the dog casts right around deep to the point of balance before coming in (see Figures 3.1–3.4). 'Distance' is also important so that the dog isn't drawn towards the sheep.

Many handlers claim that their dogs have a 'pear shaped cast', more as an excuse for the way the dog runs narrow and then breaks (to use the term loosely) around them at the end. This is not real casting. Also the shape of such casts is more like a question mark, or like a very long narrow pear, rather than a real pear.

The good casting dog should cast fairly wide right from the start of the cast, but then break out even further if it spots sheep further out.

Because real casting ability depends on a number of traits, a dog with poor casting ability will also be deficient in one or more of these traits. If the dog isn't a good casting dog, it cannot be a good dog full stop despite some people saying we no longer need good casting dogs because all of the casting is done on the bike).

A dog that doesn't cast well will not work well either.

The 'blind' searching cast

This natural, wide, sighted cast is mostly separate to the 'blind' searching cast. Even if no sheep are to be seen, the good blind casting dog tends to just keep casting out until it finds sheep, whereas most dogs go so far and then cut across, or hesitate (see Figure 5.1).

Yet some overlap between the blind and the searching cast does occur, when the dog that has been casting 'blind' then sees sheep, and so is then casting 'sighted'. However, it should only be casting sighted to an extent, because it may still be casting blind in relation to searching for others.

There is a difference between a real searching cast, and the dog simply running around sheep. Most dogs only do the latter. A good analogy is to consider the earth orbiting the sun. The earth is held in its orbit by its attraction to the sun. This is like a dog simply running around sheep – it is attracted to the sheep and runs around them, held in its orbit by attraction to them. The big casting dog, however, breaks right out of this orbit, out of any attraction or connection with the sheep, and casts free of their influence or attraction. This is the true searching cast.

Because the sighted cast and the blind cast are inherited separately, it is possible for a dog to have a good blind cast and yet not cast well 'sighted'. That is, you can have a pup of eight or 10 months old with the natural pull, which you can cast 200 or 300 metres around sheep that it can see, and it will go and bring the sheep back without a word of command, but it may cast dead straight. And yet that same pup, if it has the blind cast, if started off right, will break right out and cast way out wide searching for sheep, and cast around the boundary of a paddock and disappear into the distance.

You can also have the opposite – a dog that casts well on sighted casts, but not when it can't see sheep. The blind cast and the sighted cast are separate.

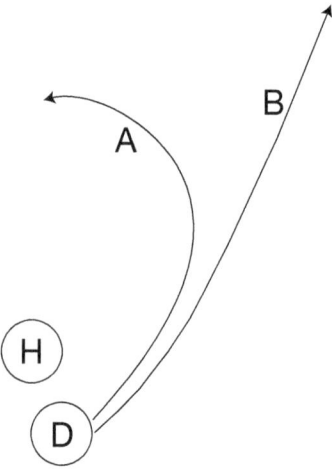

Figure 5.1 (A) Most dogs, and (B) good dogs, casting blind

Mustering ability

Even dogs with reasonable 'hold together' when working splitting sheep, and with a good cast on short, sighted casts, and a good blind cast, may not keep breaking out around stock scattered over a big area.

No matter how good a dog is otherwise, I am never really happy with it unless it possesses this ability to muster scattered stock in rough country. If it lacks this ability, it can never be trusted out of sight.

'Casting break' is the dog's ability to break out when it sees sheep further on – even if they are hundreds of metres away – in order to muster scattered sheep. The good dog always keeps out on its own side of all the sheep it can see, and never crosses between sheep.

As Michael Croft described his big casting Kelpie, Riana Nap (Scanlon's Dom × Koomalah's Stella): 'If he's casting, and sees sheep further out, he drops his shoulder and breaks out at right angles'.

A good dog always breaks out on its own side of any sheep it can see, and doesn't cross between sheep, or cross-head a breakaway, even as a pup.

When you see pups or dogs that cut in between one sheep and the others, or readily cross the cast when they can see sheep, or cross-head a breakaway, or that are quite happy to cut stragglers off the tail and leave them behind, then you can be sure, without testing them any further, that they have no real mustering ability.

Dogs that show the 'anticipating' or 'repositioning' break (mentioned in Chapter 2, under the heading 'Intelligence'), that are willing to break off sheep to get to a new position, are more likely to show good casting break.

Mustering ability is also dependent on the dog's inherited ability to look and search for sheep while it casts. The best dogs will look right and left as they cast, and even when they spot sheep will continue looking to both sides as they are casting around those they have seen.

Also the natural mustering dog has the tendency to be willing to leave sheep and go looking for others. They understand inherently that there could be more sheep about. However, some softer, wide breaking dogs can break off sheep and go looking for others whenever pressure is applied to them.

One bitch I had, Cummins Creek Chocker (Moorlands Snow × Cummins Creek Kate) with exceptional mustering ability, gave a good example of these abilities one day. I cast her down a laneway that opened into a paddock at the end of the laneway, in which a mob of goats were grazing. She ran down the laneway and through the gate into the paddock.

She then went right past the goats with hardly a glance at them, and continued on until she climbed through the far fence into a very overgrown creek. I couldn't figure out what she was doing – even from my vantage point up much higher on a hillside I couldn't see any reason for her actions.

However, after about five minutes a goat appeared through the long growth and climbed through a hole in the fence, followed by about 20 others. How

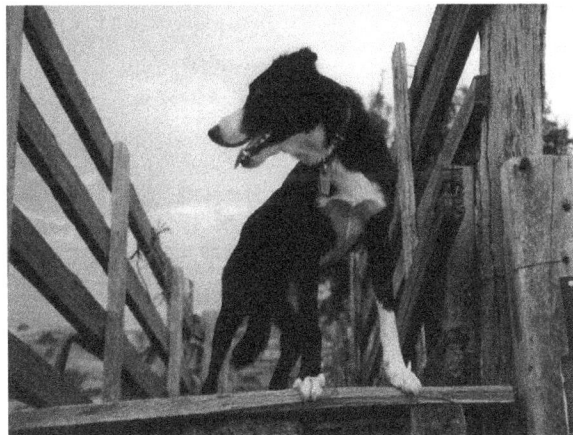

Cummins Creek Chocker (Moorlands Snow × Cummins Creek Kate)

she knew they were in there I don't know. Finally she appeared through the fence behind them, put them back with the other goats, and brought the lot home.

Another dog I had for a short time was Castleton Herbie (Bos × Castleton Judy). (Bos was by Miller's Pete 2nd, and Judy was by Cavanagh's Esjay). Vin Stapleton gave him to me as an old dog to breed from, and I worked him while I had him.

If ever I wanted to leave a sheep behind for some reason, I had a job to manage it with Herbie. I would leave the sheep and would then have to keep a close eye on the dog to prevent him from going back for it. After taking the mob half a mile away, and thinking that Herbie would have forgotten about the single sheep by now, I would take my eyes off him for a second and when I would look around again he would be disappearing into the distance, going back for the one sheep.

However, some mustering-bred dogs tend to break very wide off sheep for little reason, and turn their back on sheep and break off their work completely as they do so, and work in a fairly loose, 'scatter-brained', or 'windy' fashion. Sometimes, as soon as the handler or sheep put pressure on them they can start breaking off sheep and running around looking for others, as a way of reducing the pressure. These dogs are far from ideal.

A dog can muster in some sort of fashion for a number of different reasons. But *only if it does it for the right reasons will it be a top mustering dog*, and also be good as an all-round dog in other areas.

Listening for and scenting stock

Other aspects of real mustering ability are the instincts to listen and scent for stock. First, the dog must be calm and sensible, and willing to stop and stand in one place and look for stock, rather than just going flat out without a thought in its head.

You will see a good pup, when cast out blind, looking both ways as it casts out, and sometimes it will go so far and then stop and have a good look around. If it can't see sheep it will then break back out and continue casting (this is different to a dog that simply stops or hesitates because it is getting impatient, or is about ready to give up).

But some dogs will stand and look around, and also lift their nose and scent the wind, and listen. All (or nearly all) dogs can see, and hear, and smell; but very few use these senses to anywhere near their potential. Very few look for stock in the far distance, or listen, or notice scents on the wind (or on the ground), or react to them, either when they're standing still or when they're on the run. But these are all inherited characteristics of the best mustering dogs.

I have a bitch now that – if sheep disappear in the distance over a rocky or bracken-covered hillside, and she is sent after them (not having seen them go) – will get on their trail with her nose to the ground. She will follow the scent and disappear out of sight over the hill. Perhaps five minutes later she will return with the sheep.

Another of my dogs, Campaspe Bounce, is only young, but I have seen him track a mob of sheep that have gone through a gate 10 minutes earlier, and he will not deviate from their path until he sights them. If a dog has this instinct, it is not hard to train it to 'find' stock to command.

Others will react to scents without such an obvious display – you will see them cross the path sheep have taken and double back to go after them.

You can sometimes see signs of such abilities in pups. For example, they might stick their nose in the air and scent the wind, or follow scents on the ground, or you will notice that they can see sheep a long way away on a hillside. But once again, without a calm temperament, 'distance', and minimal eye, such abilities will never surface when at work.

Scope

A dog with a great deal of scope is willing to cover a lot of ground, with no hesitation or impatience. They make any given area look smaller than it is, while a dog without much scope (what I call a 'small-scale' worker) makes even a small area look too big for it to handle. The dog with scope is willing to cast long distances without hesitation, without needing constant urging. It just goes and goes and goes.

Most Australian dogs lack this real scope, whereas many UK dogs have a high degree of it. Most modern Australian dogs will cast only so far (and not very far at that), and then cut across or look back, and need further commands. 'Am I there yet?!' is a very good description of such dogs. They are too impatient and give up quickly.

The dog with scope has great patience, and is willing to go a long way and take a long time to get somewhere. If it also has good drive and distance it will be willing to work way off wild sheep, even though it may be very strong.

Complete casting

The pup with a natural pull, and a natural wide sighted cast, and distance, can be sent around quiet sheep some distance away, say 50 or 100 metres, and will go around them wide and deep and work them back to the handler with very little training in casting at all. This is the basic natural cast, with the natural pull.

With such a pup you can increase the distance that it can cast in leaps and bounds, especially if it also has the natural blind cast (because as the distance increases it can no longer see the sheep when it sets off).

For example Moorlands Georgie would cast a kilometre with ease before she was 12 months old, wide and deep, and muster a paddock on her own at that age. These abilities all disappear when people are breeding for trial work or yard work.

Summary

In summary, we have four main areas of casting ability (the natural pull, the sighted cast, the blind searching cast, and mustering ability), which can be further broken down into other traits which they are made up of, or affected by.

If you can get all of these things in a dog you will have a top natural casting and mustering dog with very little training at all. They are very rare.

6

The high-calibre working dog in summary

As you will have seen from the earlier chapters, the top dog's ability is made up of many seemingly contradictory aspects.

Opposing traits

For example 'force' must be balanced by 'distance' and by 'holding ability', and 'heading instinct' must be balanced by driving instinct, and 'eye' by no 'eye', and 'steadiness' and 'minimal activity' by speed and 'cover'.

So even though it may seem as though two traits are mutually exclusive, this is not the case with the traits as I have described them.

To be a top dog a dog must possess these opposing and balancing instincts. This is an important principle, and should be kept in mind when breeding dogs. Also remember that a calm, bold temperament is the cornerstone of the best practical working dogs.

To recap, we should look for the following attributes in a dog:

1   temperament and intelligence
2   steadiness and minimal activity
3   controlled force
4   holding ability
5   driving ability
6   cast and muster.

Temperament and intelligence

- Calm
- Bold

- Intelligent
- Friendly

Steadiness, minimal activity, and the natural stop

- Steadiness, work at a walk or trot
- Affected by calmness and drive
- Minimal activity; doesn't move one more step than required
- Natural stop (pause on point of balance or head)

Controlled force

- Strength (absence of fear)
- Desire to force
- Calm, controlled bark
- Controlled bite (snap head or heel only, no hanging on)
- Control of force (calmness + strength + drive)
  - Should force directly

Holding ability

- Heading instinct
- Freedom of movement (influenced by temperament)
  - Gets right around in front of head and behind tail
- Width (keep some distance out from stock as it curves around them, made perfect by square break and freedom of movement)
- Hold together (desire to keep all sheep together)
- Break
  - Breakaway break (break wide off sheep running away from dog)
  - Square break (turn squarely before moving around sheep)
  - Pressure break undesirable
- Cover
  - Short cover
  - Mob cover (width + hold together + break + freedom of movement (get to head) + short cover)
  - Square cover (includes square break and short cover)
- Eye
  - Free eye, not pressure eye
  - None most of the time
- The natural stop

Driving ability

- Drive

- Distance (drive + calmness + square cover + minimal activity + natural stop)
- Balance (trait in its own right, influenced by drive + heading instinct + square cover + minimal activity + free eye + natural stop)

Cast and muster

- The natural pull
  - Needs driving ability and distance
- The shape of the sighted cast
  - Affected by square break, width, breakaway break, scope, freedom of movement, casting break
- The blind searching cast
  - Affected by scope and break
- Mustering ability
  - Casting break (to break and recast around scattered stock)
  - Dog must always keep on its own side of any sheep in sight
  - Search for stock while casting – look, listen, scent
  - Hold together
- Natural lift (dependent on the natural stop + distance)

Physical attributes

- Agile and athletic
- Very fast and smooth
- Good natural muscle tone
- Stamina

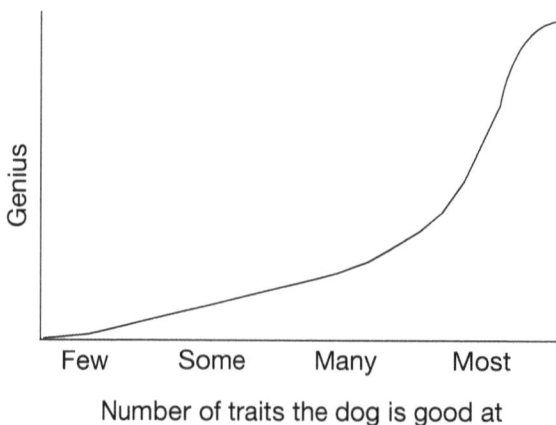

Figure 6.1 Exponential increase in dog's genius as ability improves

The best, and the rest

There is a large gap between the ability of the very best dogs, and the rest. This is because one fault is often enough to ruin a dog, or greatly reduce its total ability. There is often a fine line between a top dog and one not worth bothering with. For example a dog can have everything, and yet lack force, and so is almost useless. Or it can have everything, but is too excitable, and so this greatly detracts from its usefulness.

As dogs possess more desirable traits, and fewer faults, so their total ability gradually rises. But as they get close to having all of these traits in high degree, and no real faults, so their total ability skyrockets, and you see the real 'genius' of the best dogs. This applies not only to the dog's total ability, but also to its ability in any area, such as holding ability, or mustering ability.

The above graph (Figure 6.1) illustrates this idea.

Summary

Although I have tried to break the dog's ability down into individual traits, the situation is extremely complex and variable, and almost impossible to categorise neatly. *The basic principle is that the dog should do everything on instinct in the most effective manner, without much training at all.*

It should hold and cover wild breaking sheep, and cast long distances blind, and naturally muster scattered sheep, and naturally draw (drive) sheep to the handler without looking to head them, and happily plod along quietly behind the tail of a mob, and have the strength to confidently and calmly force a single ram, or a cow with a calf, bark when required (but never out of weakness or excitability), and not take a step more than is necessary to achieve all of this.

Don't make excuses for a pup because it is 'only a pup'. If it has the necessary speed to handle sheep, then it should handle them. If it can't, it lacks the right inherited instincts, and *they will never improve*. You may hide faults with training, in dogs kept for trial work or on small farms, but not in hard practical work. And you can't hide the faults when breeding.

Look for the old dog in the young pup.

7

The case for the 'all-rounder'

There have long been disputes over the merits of specialist dogs versus the 'all-rounder' or 'utility' dogs. Some argue that an 'all-round' dog will only ever be a 'jack of all trades, and master of none'. However, as far as I am concerned, there is only one possible good dog, and that is the all-round dog. (However, what many people describe as all-round dogs are often little more than *yard* dogs with some *slight* degree of ability in the paddock.)

To be a specialist implies that the dog lacks in certain areas that would allow it to work outside of its specialty, or that it has certain faults that would prevent the same.

For example, to describe a dog as a 'specialist paddock dog' may imply that it is a wide working dog, with some 'eye'. But the reason that it is a specialist is because it lacks the 'force' necessary to be good in the yards. Yet if that is the case then it lacks the force necessary to put sheep over a creek, or to force reluctant sheep on a hot day, or to put a big mob into a difficult set of yards, or the strength to handle cows and calves, and so on. So a specialist paddock dog is simply a second-rate paddock dog.

The same applies in reverse to the 'specialist yard dog'. The reason they are no good out in the paddock is probably because they split sheep up and lack 'holding ability', their temperaments are too excitable and ratty, and they generally lack much ability. But all of these shortcomings are detrimental in the yard just as they are in the paddock (more so in a decent set of yards than in the small yards of a yard dog trial). So a specialist yard dog is simply a second-rate yard dog.

Of course you can get away with such a specialist yard dog in some situations, if you want to, the same as you can get away with a specialist paddock dog, but neither is ideal.

Also if you have been out mustering sheep in rough country into a set of out-yards to drench them, for example, you can't then go home for your 'yard dog'. Your mustering dog must do the yard work also (it can never be the other way around, because your specialist yard dog hasn't a hope out mustering). And if a few sheep jump over the fence of the out-yards and bolt down the hill for the nearest creek, your yard dog isn't going to have a hope of getting them back.

From all accounts, the *best* of the original Border Collies and Kelpies in this country (and the old collies in the UK) had great casting ability and 'distance', and also the bark and force when required, and the right temperament (calm and strong) to use eye only when needed. They were all-round dogs. There must have been great variation in the dogs then, the same as there is today, but the best dogs from all accounts were exceptional.

In more recent times, for example, Jack Hiscock's best dogs of his old blood would work cattle or sheep, paddock or yard, big mobs or small, and win big three-sheep and field trials besides (with a top handler). And yet he talked of dogs of years earlier, such as those owned by Sinclair (some of which were sourced from James Moore, and had Macpherson's Moss blood close up), as being better again.

In my view the ideal to aim for is the all-round dog. Anything else implies limitations and shortcomings.

8

When things are not as they seem

Evaluating sheep dogs is very difficult, and only a handful of people, after years of experience and thought, can accurately assess a dog's (or pup's) potential with any degree of accuracy.

This is partly due to the fact that there is so much overlap between traits, and that some traits can be confused for others, and that temperament affects the expression of various traits.

I hope that the following pointers may help to explain why many pups do not turn out as expected, and may help readers choose their breeding animals wisely.

When the breeder looks at a litter of pups and thinks, 'But they're not what I thought they would be!', most likely the fault lies not with the dog and bitch not having bred as their ability suggests, but with the breeder because they did not understand the dogs' abilities. Often the breeder has not understood the traits that make up the good dog, nor the ways that the traits and the temperament interact.

Also when people say that a dog was a top dog, but never bred anything, I am more inclined to think either that their assessment of the dog was inaccurate, or that the dog was never bred to any decent bitches of a similar type.

Often the progeny of a dog can give you insights into the parent. A pup may show a particular trait and yet you have never noticed that trait in the parents before. But when working the parents later, or seeing them work (perhaps in a more difficult situation), and having become more attuned to that trait, you may realise that it was there all the time only in a more limited fashion, or perhaps hidden by some other trait. And so you realise that it was your evaluation of the dog that was at fault.

Often the main reason mistakes are made is that the dog is not seen tested in a variety of challenging situations. It is difficult to evaluate a dog on limited information, but the more you see it tested to its limits and beyond, the more accurate you will be. For example, you can see a dog work at a trial on easy sheep under mechanical control, and it may look okay. Another day you can see it on difficult sheep and realise that it isn't worth the price of its feed.

Interactions between traits, and between traits and temperament

There are three main ways that traits and/or temperament can confuse the evaluation of working dogs.

First, some traits can be mistaken for other traits, that is they can give the illusion of inherited ability that doesn't really exist. I call these 'counterfeit traits'. Just as counterfeit money may look like the real thing on the surface, and yet on closer examination be revealed as worthless, so some traits may look like the real thing in easy situations, and yet when tested in difficult conditions prove not to be.

Second, some traits can hide faults in easy situations. I call this the 'masking of faults'. There is considerable overlap between this point and the previous one.

And third, temperament – either excitability or weakness or tension – can modify the way many traits are expressed, as can physical ability. I call this the 'modification of traits'.

So we have:

- counterfeit traits
- the masking of faults
- the modification of traits.

An understanding of these points will help to determine more accurately what a dog's inheritance is, and so how it will breed, and also how it will perform in difficult situations.

Counterfeit traits and the masking of faults

The three main traits that I believe can give a dog the appearance of possessing certain abilities without actually having them, are certain forms of eye, break, and weakness (especially in combination, which they often are).

For example a lot of eye can make a dog seem to have 'minimal activity', or the 'natural stop'. But if you took the eye away (or worked the dog on the tail of a mob where it didn't use any eye) it may be as busy as they come. Thus eye can produce a 'counterfeit natural stop', or 'counterfeit minimal activity', or can hide busyness.

And if the eye is largely due to (or exacerbated by) weakness, which it often is, and you breed the bitch to a strong, steady dog lacking in eye, you can get very busy pups with no natural stop and no minimal activity. You look at the two parents and think, 'Neither of them are busy, where did that come from? Must be a throwback,' when in actual fact the bitch was extremely busy; it was just that the busyness was hidden behind excessive eye. The eye gave a false impression of minimal activity or the natural stop.

Weakness can also hide faults. For example, you may think that a dog has good 'width' or 'hold together', and doesn't split sheep up. However, this may be due to the dog's weakness keeping it out off the sheep in easy situations, not that the dog really has the desire to keep stock together. So weakness can produce a counterfeit break, width, hold together, or cast, and yet in fact the dog may possess none of those abilities in their true form.

Eye can do similar things. The eye steadies the dog and makes it stand and stare so that it doesn't split sheep. But if you took the eye away, the dog might split sheep all over the place, as it might lack the real hold together and width and cover that a good dog must possess. You would see these faults come out when working more scattered or difficult sheep, and when breeding. So eye can be confused with other abilities, and can hide these various faults, *in easy situations*.

Break (as desirable as it is) can also make a dog appear to have this width or hold together to keep out around all of the sheep. If you only see it worked in a small paddock, it may seem that way. The dog breaks out around all of the stock, and musters them all.

However, generally it only does so when the distance between animals is less than its 'break zone'. If the sheep are further apart than that then it will miss some and leave them behind. The more tense, excitable or weak this type of dog is, the wider it will break, and so the more of this casting and/or mustering ability it may seem to possess.

'Pressure break' can also give the illusion of width and of casting ability, when the dog possesses neither of those, and the dog will breed that way. Breed the dog to a similar type, that is; weak with this pressure break, and you will get pups with that trait because they also have the weakness. But try to improve the strength in the pups by joining the bitch to a strong dog, hoping for the wide working ability of the bitch with the strength of the dog, and you are doomed to failure. Of course the right sort of break is desirable to have in any dog, but you must realise that break can appear to be things it isn't, and can hide other shortcomings.

Also, in many dogs break is linked with eye, and the dog only breaks out when using eye – if working plainly many will go straight through the middle of sheep, because they lack width and hold together.

People also often confuse pushiness for strength. Many pushy dogs are not very strong at all, which becomes obvious when something really has a go at them.

So usually when pups don't turn out as expected it is not that the parents haven't bred true. Rather it is because the breeder hasn't understood the fact that certain traits can appear to be other traits, or can hide faults, and so hasn't discerned the parents' inherited make-up accurately.

These are just a few examples of the ways in which some traits can hide shortcomings in easy situations, and give the impression that a dog possesses characteristics that in fact it doesn't. But if you push the dog to its limits you will see these traits for what they really are, and see what lies beneath.

In my experience most dogs breed very closely to the way they are, when you understand exactly what their make-up is. Of course, chance is involved and some pups will get some traits more strongly than others, but basically they will be various fairly predictable combinations of their parents' characteristics.

The modification of the expression of traits

Certain traits modify the expression of other traits, and this has been touched on above. This too can confuse the evaluation of dogs. The greatest single way that this occurs is through temperament – that is, strength or weakness, boldness or timidity, calmness or excitability (nerve or tension).

A weak or tense dog will generally show more eye (particularly when under pressure), and will often show more break and thus a sort of cast. A strong dog, on the other hand, will tend to show less pressure eye and less pressure break for any given amount of a certain type of inherited ability. (However, it is entirely possible – and essential – to have very strong dogs with good break.)

Also, the amount of eye that a dog uses or shows is often very dependent on how calm or otherwise it is. The dog that will show no eye on a mob and happily plod along droving without a care in the world (which is what is required), and then tighten up and show eye and great balance and cover on wild sheep, is the calm, relaxed dog.

Another factor that can modify the expression of a dog's traits is its physical ability. For example, a dog may be a steady worker simply because it is poor physically, and is slow rather than steady. It may also appear to be a calm worker for the same reason, because it simply doesn't feel up to rushing about.

I remember one bitch that was extremely fast and excitable, and one day she turned suddenly on hard ground and tore her front pads. Immediately she looked a million dollars, showing great balance and short cover. She was a very good bitch as far as her inherited traits went, but her excitability ruined her. Temperament and work are inseparable.

Also a poor moving dog will not be able to show a high level of holding ability when working wild sheep, because it simply lacks the physical ability to stop and turn quickly and lightly while moving fast.

So be careful when breeding for steadiness and so on that you aren't simply breeding for slow moving dogs, or that when breeding for cover that you aren't breeding for excitable or busy dogs.

And remember that temperament and work are inseparable. The expression of many traits is greatly affected by temperament.

Stylish workers and plain

It is easy to take a fancy to 'flashy' workers – those with eye and style. Plainer workers are often overlooked in favour of such dogs. Yet much style stems from weakness or tension or excitability, and is often counterproductive.

When talking about how style can deceive, it is useful to categorise working dogs into three categories: poor dogs, handy dogs, and top dogs (although in practice there is continuous variation between each category). Very poor dogs make any job look very difficult, and lack the ability to be able to handle most situations, and so people can see them for what they really are: poor dogs.

Handy dogs, such as many stylish dogs, make the job look hard and make more work for themselves, but have enough ability (when handled well) to cope with the situation. Because they make the job look hard, and yet succeed (more or less) in a showy fashion, people are deceived into thinking that they are top dogs. *People confuse style for natural ability.*

Finally, the top dogs are those dogs that are so good that they make the job look easy, and in many situations don't appear to be doing much at all, and so they can get passed over in favour of the flashy dogs which look to be doing a lot, but aren't even in the same class.

Don't be fooled by a flashy bit of style, unless it comes with real substance. Remember that the less effort a dog uses to get the job done, the better it is. The good dog is fairly plain in most practical situations, although it will be stylish when really needed.

Summary

In summary, I hope that these comments will help you to understand why breeding is so difficult and often disappointing, and help point you on the road to breeding better dogs. I hope it will prevent much wasted time and effort trying to produce a 'silk purse from a sow's ear'.

You must try to look beneath the surface of a dog's actions, and determine *why* it is doing what it is doing, and what traits it does, or does not, actually possess.

Is it keeping out around the sheep because it has real hold together, or just because it is weak and breaks wide? And so on, for many other traits. Only if you can answer these questions will you have an accurate idea of how the dog will breed, and how it will perform in challenging conditions.

Remember that often the plainer pup not doing much wrong is the better breeding prospect than the flashy, stylish dog that may look nice, but underneath the eye and style there is nothing. 'All style and no substance' is a highly accurate description of many such dogs.

9

A few good dogs

What I would like to do in this chapter is to describe three dogs that I have owned or worked, and break their ability down into the individual traits outlined earlier.

The three dogs that I have chosen are not perfect, but they are (or were) exceptional in many regards. These descriptions might help to show the ability of the good dog to those who have never had the good fortune of owning one, or had the chance of testing it in challenging conditions.

Boho Misty (Marlowe Clyde × Native Point Gwen)

Boho Misty was a yellow-and-white Border Collie bitch I bought when she was 12 months old, broke in and trialled, and worked at station work on sheep and cattle. She was exceptional in many areas.

Temperament

She was exceptionally calm, very bold, and incredibly intelligent. When training her I only had to show her something once – sometimes it seemed as though I didn't even have to do that. The second time I asked her to do it, she had it down pat. I remember particulary teaching her to break out 'get out' on the cast. I attribute a lot of this intelligence and trainability to the UK blood on her mother's side.

Boho Misty

She was exceptionally loyal and protective. For example, she didn't like vets (in fact one day she had the vet bailed up against a wall, working her with eye and style) and if I was walking around the property with the vet, Misty would very deliberately position herself between me and the vet, and keep a close eye on her.

Holding ability

She also had very good 'holding ability'. She possessed perfect 'heading instinct', and would get right around in front of the head even though she was a natural driving dog.

I have never seen a dog with a greater ability to hold one single wild sheep up to the handler in the open, and hold it completely immobile. A handy dog can generally work one sheep up to the handler, but the sheep tends to then go around and around the handler, with the dog going round and round the outside. Not with Misty.

Her ability in this regard was due not only to good square cover, and drive, and getting completely to the head, but what set her apart from other dogs was her distance and minimal activity and natural stop. The instant that she felt the sheep was balanced and ready to stand still she would stop dead and maintain the situation. Most dogs move just that fraction and so start the sheep moving again.

She was completely free in all her movements around stock. She had good width and would give them good clearance as she curved around them, more so the wilder they were.

Her break gives a good idea of the complexities of this trait. She had exceptional square cover and square break, and would always keep her distance, and break off wild sheep, but if one sheep broke away she would shoulder it to head it (she wouldn't bite). She lacked breakaway break.

In fact, in relation to break, another example of her intelligence is that she later learnt to break when heading a breakaway. Once, when yarding a mob of sheep, one broke away down a fence line. Misty took off after it and was making as if to shoulder it, but there was a big tree in her way so she broke out around the tree and headed the sheep cleanly. From then on she mostly used to break out when heading a breakaway.

She had good short cover – good, steady cover, not the jumpy, windy sort of work some people admire.

And she had exactly the right sort of free eye. She would use it on the tail or any other part of sheep, with no stickiness. Very stylish at times, but no stickiness or doughiness. She would also use it to work other animals, including rabbits or chickens, and even small lap-dogs. I could work a small dog with

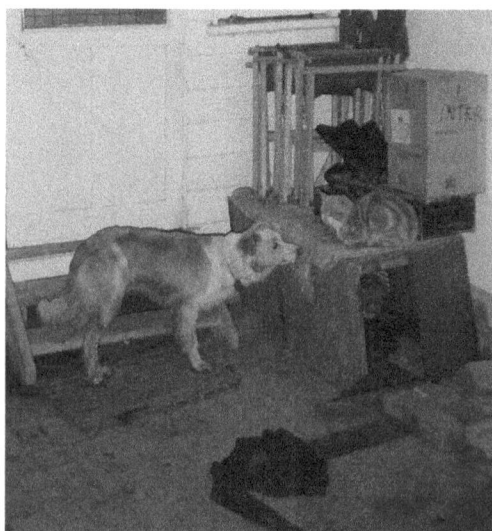

Boho Misty keeping an eye on the cat

Misty as if I was working a sheep (although she didn't work other dogs as a rule, as some dogs do).

She would also work ducks on a dam, and obey the directional commands while swimming.

Drive

As for drive, again she excelled in this area. Given her great driving ability – and exceptional calmness, natural square cover and square break – she had great distance, better than any I have seen. When working wild sheep she would work naturally at a great distance, and as she got them broken in she would gradually work closer and closer.

Her 'balance' was also very good. She also had a very high level of direct walk-in, and with her you could apply force at any point with great precision.

Cast and muster

She was a good casting dog with good scope, but lacked the mustering ability that would have made her a top casting dog.

She was very easy to teach to break out in response to 'get out'. She had a perfect natural lift, and always cast to the point of balance, never stopping short.

She didn't have the real natural mustering ability, as she didn't have the idea of searching for sheep or breaking out around sheep hundreds of metres further out. She was no slouch in normal situations, and would muster cleanly in relatively simple situations, but wasn't a natural out in rough country.

She would recast to command easily, due to her distance and temperament and scope.

Controlled force

Boho Misty was a strong bitch, in fact it got her into trouble a number of times because she would happily squeeze around the feet of cattle, or around goats in the yards, and sometimes got kicked or rammed for her efforts.

In fact, one day she stood up to a goat that charged and hit her head-on, almost knocking her unconscious, and it was this incident that taught her to heel the front foot. As I have said before, she learnt like lightning, and could put two-and-two together in an amazing fashion. The next time a goat charged she still didn't show any weakness, but just waited for it and then, as cool as you please, ducked past its head and bit its front foot. She had never done this in the past, on sheep or cattle or anything.

But next time working cattle she adopted exactly the same technique; no aggression or fizz, just cool and calculated – controlled force. There is nothing more effective on charging cattle than the front foot bite.

She would snap at the nose and bite the front foot. She wouldn't heel the back foot though, and she wouldn't bark.

Steadiness and minimal activity

She was a very steady dog when nothing needed doing – walking when that was the case, and she had great minimal activity. There was not a busy bone in her body; she always did as much as was required, but not a step more. She was completely free moving, and would speed up as much as the situation demanded.

So, in short, she was an exceptional dog. To be perfect she needed the breakaway break, more mustering ability and some bark.

Moorlands Georgie (Moorlands Snow × Moorlands Emma)

I bought Moorlands Georgie at about 10 months old, trained her, and worked her on a number of hill country properties including 'Shepherds Creek', which was about 10 000 acres running approximately 13 000 merinos and some cattle.

Both of Georgie's parents were by Cavanagh's Esjay (a dog bred by Sid Cavanagh, which Jack Hiscock rated as the best all-round dog he'd worked), and she had five crosses of Yulong Russ fairly close up (see her pedigree in Chapter 11).

Temperament

Georgie's temperament was not ideal. She was a bit excitable, very hard (not that I mind that), and of only average intelligence (although she was a much 'brainier' worker than most).

The combination of exceptional cover and being somewhat busy and excitable, with a very hard temperament, made her extremely difficult to train and to gain control of. However, her natural ability made the effort worthwhile.

Moorlands Georgie – a hard, tough, self-reliant worker of exceptional natural ability

Holding ability

She was an exceptional holding dog, and had great 'width' and 'hold together' and 'casting break', and would never split sheep. Even with scattered sheep in big paddocks in rough country, she would rarely miss any. She also had excellent 'break', and always arced around her sheep in a nice curve.

Due to a touch of excitability and busyness, and her forceful nature, she tended to get in on top of her sheep when holding them. She would get away with it because of her great 'cover' and 'block and hold' and 'break', but would put much more pressure on a small number of sheep than is desirable.

As already stated she had exceptional cover, both 'short cover' and 'mob cover'. I have never seen a dog with more cover. Combine this with her break and width and hold together, and I have seen her do some amazing things holding wild, fast breaking merinos out in rough country. One particular incident (among many others) involved two big merino lambs that had been missed in a muster. I was sent out to see if I could find and yard them into a set of out-yards.

After eventually finding them in a big, rough paddock, we (Georgie and I) worked them back through the stringybark and box timber to a set of out-yards in the corner of four paddocks. The ground was rising up to the yards. As we got within a couple of hundred metres of the yards I went ahead to open the gates.

I stood up at the yards and watched Georgie in action. So not only were the yards in the corner, which made the lambs feel trapped and so reluctant to come up, but I was also standing there, and the ground was rising which gave the lambs the advantage on Georgie.

Anyhow, as the two lambs got within about 150 metres of the yards, they made a simultaneous break in opposite directions. Georgie broke back out and headed one, and once it was turned she left it to head the other one (all this was completely without command of any sort).

(She had an amazing ability when mustering, or working split sheep, to keep a picture in her mind of exactly where every sheep was, and what was happening around her. She could be working one lot of sheep and as soon as they were under control she would leave them to go and retrieve some other situation that needed attention.)

Anyhow these two lambs were about 50 metres apart, and trying to get away. But Georgie would move one in towards the yards and towards the other lamb, then, judging her moment, leave it and do the same with the other. Over about five minutes she gradually put them back together with the most extraordinary display of cover and break and hold together I have ever seen. She then pushed them up the hill and into the yards, all without a word being said.

At lamb-marking times often a few lambs would get out of the yards and disappear over some hill and out of sight. At a word Georgie would break out over the hill and disappear after them. I remember one particular occasion when the marking contractor was giving me strange looks when this occurred after half a dozen or so lambs got out and disappeared, and he kept looking over his shoulder to see what the outcome would be. After a while Georgie returned over the hill with the six lambs under perfect control (making it look ridiculously easy), whereupon I opened the gate and she put them back into the yards, all without a single word of command.

She would also nudge a young lamb in the required direction with her nose if it was a bit slow on the uptake, as would Cummins Creek Chocker (Chocker was more than Georgie's half-sister – by Moorlands Snow and related on her mother's side also). Chocker would do the same thing when working chooks.

I have also seen Georgie covering sheep in midair. When working on the ground she would rarely overrun a step, covering with great speed and precision. I have often seen her when a big strong wether has attempted to leap over her, leap into the air in front of the wether herself, and as it tried to twist and turn in the air Georgie would also twist and turn with great precision, covering every slightest movement, even though both she and the sheep were completely airborne. When the wether's feet hit the ground it was beaten, and would dive back into the mob.

Eye

She had the right sort of 'eye', as some dogs of this blood do – none most of the time, but enough when needed. And even when using eye she would still be watching the whole mob.

Drive

Georgie possessed good 'drive' in her inherited make-up. She would naturally cast around to the tail of sheep from a young age, even if they were wild and in a tight corner, without trying to hold them into the corner, and was a natural pulling dog.

However, she would tend to go too far up the wings when droving a mob along, and didn't like forcing on the tail, sliding up towards the head (although she would heel at times), but she would never ring sheep on the draw right from a pup.

Distance

Because of a touch of busyness and excitability, she tended to be drawn by sheep, and would crowd them on the draw – she tended to get in on top of them. She could get away with it most times because of her exceptional holding ability, but it is not ideal.

Balance

She had exceptional 'balance' in her inherited make-up (that is she knew exactly where the point of balance was), but being somewhat busy she tended to move about more than was required. If you steadied her down with control she would balance with the very best. Once again it shows how many different traits must come together to make the perfect dog.

Cast and muster

At casting and mustering she was a rare type of dog. At 12 months old she would cast a kilometre (measured, not guessed), wide and deep, and muster a paddock on her own and work the mob back to me. She was an exceptional natural mustering dog in rough country, with a huge searching cast. Right from the start of her training you could see her looking right and left as she cast out, searching for sheep. She readily understood to leave sheep and recast, searching for more.

Some big casting, specialist mustering type dogs can break excessively wide with very little provocation, and can be fairly loose, and this is where much of the mustering ability is coming from. What was so good about Georgie (and other dogs of this line) was that she had all of the mustering ability and yet was also a very tight-holding, strong-blocking dog.

She would never go between one sheep and another. Even if you were working sheep along a road, and sent her to head the mob, she would always squeeze between the fence and the outside sheep, even if there was a big opening inside that which would allow her to get to the lead more easily. She wouldn't split sheep. Nor do I recall her ever crossing once she had seen sheep, in all the years I worked her, but I suppose she must have done so at some time.

She was also totally reliable working out of sight. You could be droving a big mob of ewes and lambs or something for miles, or mustering them, in rough, hilly, timbered country, and might perhaps only be able to see a third of the mob, the rest being out of sight on each side.

She would work away on her own where I couldn't see her, and you knew she would never leave sheep behind. She had good drive so that she wouldn't head the mob, and could be trusted to keep them going in the right direction. After quarter of an hour or so she might appear on a ridge to see where you were, and then disappear again.

You can't leave this sort of work to many dogs. First, an over-heading dog (for whatever reason) will disappear around the front of the mob before long. Also the dog must have hold together and width and mustering ability, and casting break, so that you know that it isn't going to leave sheep behind. And most dogs don't like working on their own – they need to have a strong independent streak in them to be good at this class of work.

A kangaroo jumps a fence at dusk

Speaking of casting, on one occasion I wanted a photo of a kangaroo jumping a fence. So I found a mob lying down resting, half a kilometre away from a fence, positioned myself near the fence behind a tree, and cast Georgie out. A dog that cast too straight, or came in too early, would have simply hunted them away, but the 'roos didn't see her until she lifted them from the far side, and she brought them straight up to me and I got the photo.

Controlled force

Georgie's force left a bit to be desired. She would block any sheep alive and nothing would ever get past her, but her excitement meant that she tended to force in a 'fizzy' manner instead of steady and controlled. She was adequate but certainly not perfect. She wouldn't bark.

However, she was a top shed dog, penning up. She mostly worked on the ground which is the best place to be usually, and knew exactly which sheep to concentrate on to get the lead moving into the catching pen, and which sheep to concentrate on to keep them flowing once they started.

She would back when needed, but wasn't a natural. Her exceptional cover and block and hold, that is the ability to force and hold at the same time, is what made her an exceptional shed dog. She would force exactly the right sheep to push the one in front of it into the pen, and then move back to send the next sheep up and so on. Never a bark, never a sound, just exceptionally effective.

I have on occasions handled mobs of up to 1500 dry sheep myself in the paddock, with Georgie alone. I have put mobs (800 or 900) over creeks, with only Georgie, and myself on foot, due to her exceptional cover and her ability to force at the right place, as she did in the shed.

This is something that the handler materially assists in teaching a dog; when you have control you can cause a dog to force in a certain manner and in a certain place, and if you do it often enough the dog learns to force in that fashion on its own, without command.

Steadiness and minimal activity

Georgie was reasonably steady, but also a bit busy. With training to 'steady' to slow her right down she would balance sheep with the best, but left to her own devices was inclined to be impatient and move off the point of balance.

She was a very tough dog physically and mentally, with great heart and stamina. To be perfect she needed to be more relaxed with more minimal activity (which would have improved her distance), more outright strength, direct walk-in, and some bark. (See the colour photo section for photos of Georgie at work, mustering in hilly country.)

Moorlands Tomie (Cavanagh's Esjay × Barravore Jean)

Tomie is a black, white-and-tan short-haired collie, marked almost identically to Georgie. He is out of the same litter as Georgie's mother, Moorlands Emma.

I have never owned Tomie, but have often borrowed him from the Hiscocks (his owners) and taken him home to join bitches to, and have tried him out on sheep.

Temperament

Tomie is of exceptional temperament, very calm and bold, highly intelligent, and friendly.

Holding ability

Tomie has great holding ability, like Georgie. He has good 'hold together', and doesn't split sheep. His 'width' isn't quite right, because he has good width on the shoulder and head, but not on the tail.

He has exceptional 'breakaway break', as do most of his pups. When covering wild sheep he tightens right up, and shows great 'square cover'. His 'short cover' (and 'mob cover') is exceptional.

Tomie has the perfect type of 'eye', showing none at all on a mob or quiet sheep, looking very plain, and very stylish on a few wild sheep. There is no stickiness.

He continues to breathe calmly even when holding a couple of wild sheep and showing a lot of eye; there is no tension in him. His tail hangs straight down completely relaxed at all times.

Drive

Tomie has a great deal of natural mob 'drive' (though not the direct drive in the same class as Boho Misty or a UK dog), and this combined with his calm nature gives him good distance. He also has great 'balance'.

Moorlands Tomie – 'brains'. Relaxed concentration at its best

Tomie, given his driving ability and great intelligence and calmness, is always doing what is necessary to get a job done. Put him around sheep and he will take them to what he thinks is the most likely gate for them to go through. He has exceptional purpose.

If you cast him around a paddock from a gateway, even if you then move off to a completely different position, he won't take the sheep to you but will put them through the gateway from which he was cast, even if that means taking them away from you.

I saw this dog do a very 'brainy' piece of sheep dog work. I had borrowed him to join another bitch to, and one evening after the day's work I let him and a few other dogs off, to go for a walk along the creek. The paddock I was in was about 200 acres, not huge, but big enough in that sort of country considering what Tom achieved. The creek divided the paddock into about one-third on one side of it, and two-thirds on the other.

The country rose up very steeply on the far side of the creek, some of it too steep to ride a horse up or down, and was then hilly behind that. There was scattered box and stringybark timber over the paddock.

Anyhow I walked along this creek to the crossing place, which was simply a spot where it was a bit shallower than elsewhere with a pebbly bottom where you could drive a vehicle over. We went over this ford, getting wet feet on the way, and continued along the other side still heading in the same direction.

As we neared the far boundary of the paddock, a mob of about 400 ewes saw us and took off over the hill out of sight. Tomie decided to get them, so off he went. Not being my dog, I couldn't call him back. I decided I had better go after him (on foot still), as there was a road on one boundary of the paddock and I was worried he might get run over.

So I climbed up over the hill and kept walking towards the back of the paddock where I expected the sheep to have made for. Getting in sight of the back corner of the paddock (by this time I was soaking wet, because it had started raining) there were no sheep or dog to be seen. I thought he had probably taken them back to where I was when he cast off, and that I had missed seeing them going around the side of a hill or something.

So I walked back by a slightly different route to scour more of the paddock, to see if that was the case. No sheep to be seen. I then decided to check the one corner of the paddock that side of the creek that I hadn't yet seen, with the same result. Now I decided that I had better go home and get a vehicle, and have a drive around to see if I could find him, so I headed back for the creek crossing.

As I came over the steep hill on the creek bank, there was Tom down below with the mob held at the creek crossing (there was too much water to put the mob across without a lot of trouble). I thought I would sit down (in the rain) and watch him for a while, just to see what he did. So the four other dogs and myself did just that. Any time the mob moved away from the creek crossing he would put them back.

Now you must realise that there was really nothing much to signify that this crossing place was any different from any other point on the creek, except the vehicle tracks leading into and out of the creek on each side. Also Tom wasn't my dog and had never been in this paddock in his life. In fact more than 99.9 per cent of dogs could work in that paddock all their life and never reach this level, simply because they lack the right instincts and temperament and intelligence.

I sat and watched him for half an hour, in the rain, and he tried to put them over that crossing, but without success. No dog on its own could have managed it, but I just wanted to see what he would do, whether it was a coincidence or not. Then I went down to him, and with the help of the other dogs I had with me, and Tom, put the sheep over. I thought he deserved that much!

For a dog to find that one spot on a kilometre of creek frontage, with a mob of sheep, after taking them across a rough paddock in which you couldn't see very far, and then hold them (and put them back) on that one spot, is amazing. Yet knowing his propensity and brains to always take mobs to the gateway he thinks most likely, I am convinced that it was simply a display of the ideal natural ability (particularly drive and 'minimal eye') combined with great calmness and exceptional intelligence, and the natural pull, plus experience.

(In the section on training, on teaching to drive, I give ways that you can encourage this type of real 'purpose' and intelligence in a dog's work, provided that the dog has the necessary ability.)

Georgie often amazed me with her 'homing' instinct – or sense of direction or location – similar to this, which when combined with the 'natural pull' is amazing, and is something that is found in the best mustering dogs. I could send her out to muster a rough paddock, and she would maybe go over a creek, up and down a few hills, mustering scattered sheep and putting them into a mob, all in a metre of dry phalaris to limit her vision, and then work them directly back to me maybe 20 minutes later. And I wouldn't have said a word to her the whole time.

How she could remember where I was I have no idea, but she always did. It is hard enough for a person in a big, rough, hilly paddock they have never been in before, to remember where the gate is, but the very best mustering dogs seem to have an amazing sense of direction. I have sometimes lost lesser dogs and spent hours looking for them when mustering in rough country, because they lacked this ability.

It is this sort of instinct that enabled Tomie, in a paddock he had never been in before in his life, to find a single spot with a fair sized mob of sheep.

And then you go to a dog trial and see dogs that can't find three sheep on a cricket ground, or manage to split them up and lose them and all sorts of carry on. And handlers who complain about 'difficult pick ups' where the sheep get back on the fence and the dog must get them off, or complain about casts any longer than a football ground as being too long, or about the sheep being too wild!

The gulf between the majority of dogs being bred and the real working dog is so enormous it is ridiculous. Such dogs are not fit to be spoken of in the same breath as the real sheep dog.

Cast and muster

Tomie is a good casting (and recasting) dog, again with a true searching cast, and a good mustering dog.

I have heard from Jack Hiscock that when bringing the dairy cows in, if one gives trouble he might leave it behind until he gets the rest of the cows through the gate, and then turn around on his own initiative and cast back out and retrieve it when he can concentrate all his effort on it without the rest of the herd drifting off.

Georgie was similar. If she had too much on her plate at one time (and I often work fewer dogs than I should because I like to see how good they really are) she would deal with the most pressing issue first, keeping a picture in the back of her mind of everything else that was going on, and once that was under control she would go and retrieve the next most pressing situation.

These sort of mustering traits or 'stock sense' all disappear when people breed just for trial work, or for yard work.

Controlled force

Tomie's force is better than Georgie's was. He is a stronger, calmer dog. He has great cover while forcing (hold and force), and generally forces in a very calm, relaxed manner.

He is not a barker, although I have heard that he will bark to get dairy cows out of a dam on a hot day. He naturally heels.

Steadiness and inactivity

Being so calm, he is also a very steady worker. He will plod along calmly behind a mob and look the plainest dog in the world. But see him working one wild sheep and you quickly change your mind about that, which is exactly what you want. A dog *should* look fairly plain and relaxed in most situations, and only tighten up when required.

He also has very good minimal activity. Not as good as Boho Misty, but very good nonetheless.

To be perfect, Tomie could do with some more width around the tail, and perhaps a touch more real minimal activity and natural stop. Also his 'freedom of movement' isn't perfect, and he has very little bark.

Old Blood

Find me a real dog, bred down
Many lines from the great Yulong Russ

A dog of brains and ability, self-reliant
Of cover, and balance, and the searching cast
Hard dogs and tough
One good line, the last

Unfashionable now
When soft dogs, weak dogs, and brainless
Turn-tailers proliferate
The steady strength and cover despised
In ignorance, inexperience

The good dogs have thinned, died out
Still, a remnant remains, a handful
Natural dogs, and clean

Real sheep dogs not quite yet
A relic of the forgotten past
As a priceless pup demonstrates
The old blood, throbbing in its veins.

Part 2
BREEDING SHEEP DOGS

10

Dogs past and present

Over the years sheep dog breeds have undergone various changes. As conditions have changed and the type of people breeding dogs has changed, so too have the dogs. At various times, certain large-scale breeders have also noticeably altered the make-up of the breeds.

Even just in the past 10 or 15 years the deterioration in dogs competing in sheep dog trials has been obvious, and many experienced long-time breeders agree.

There are a number of reasons why modern dogs are so poor. *The main one is a lack of practical work, which results in breeders not understanding what a good dog is.* Another major reason is poorly designed dog trials, and a final reason is too much outcrossing. Each is discussed below.

Lack of practical work

The first reason for poor quality dogs is because of a lack of practical work – most dog breeders (particularly those breeding Border Collies in this country) are mainly breeding dogs for competitions, and many have no interest in breeding dogs suitable for hard, practical work.

This is less of an issue with the Kelpie, where a greater proportion of the breeders are also stockmen, but it is becoming more of an issue all the time with the increase in handlers and breeders who are not professional stockpeople. Also there are many 'breeders' breeding Kelpies mostly for the money to be made out of them, who have little idea of what a good dog is.

The demographic of trial handlers has changed from predominantly stockmen types who, although competing in trials, were still mindful of a dog's practical usefulness (as were the judges), to a demographic nowadays of more hobby handlers with little experience of real stock work.

Even those who are engaged in practical work mostly use motorbikes, and mostly work on small properties, and so can get away with lesser dogs than if they were working off horses on large properties. When horses were the main method of transport, and when properties and paddocks were bigger and less improved, much better dogs were sought.

'Necessity is the mother of invention' and it is also 'the mother of good sheep dogs'. As the average level of necessity has fallen, so too has the standard of the dogs.

Lack of understanding

The second reason for poor dogs being bred is because people don't understand what a good dog is. They breed from successful or flashy trial dogs, not understanding that often such a dog's success is due far more to training and handling than to inherited ability, or that most flashy style is due to weakness, tension, or excitement. Others breed from pushy 'yard' dogs with little real ability.

Because there are so few good dogs about, most people have never owned one, and therefore they have nothing to compare their dogs with. They may honestly believe that they have good dogs, and yet have nothing but mediocrity at best – they don't realise what is possible.

Without a good understanding of the traits that make a good dog, and a crystal clear aim in mind, good dogs can never be bred consistently.

Poorly designed dog trials

The third reason modern dogs are so poor is due to poorly designed sheep dog trials. These trials create an artificial form of 'necessity' that can take precedence in many breeders' minds. This issue is discussed in depth in Part 6.

Excessive outcrossing

Even if a breeder understands what a good dog is, and is breeding for high-quality, all-round work rather than for poorly designed three-sheep or yard trials, they still often fail to breed consistently good dogs.

One reason for this is the reliance on outcrossing. People have a dog with certain faults, and breed it to some unrelated dog that they think will remedy those faults. It rarely works. Consistent outcrossing is the surest way to breed second-rate dogs, and to produce a great deal of variation and no real consistency.

Sheep dog – the hardest animal to breed

Yet even if you understand the traits, know exactly what a good dog is, and know what you are trying to achieve, there are still many difficulties in breeding sheep dogs.

One reason already explained is the masking of faults, and counterfeit traits, which make assessing dogs difficult. This is not such an issue if you are constantly proving the dogs in hard, challenging work, because inadequacies become painfully obvious, but either way you must understand the traits themselves fully.

Taking cows and calves down a hill

The major difficulty in breeding sheep dogs is simply because there are so many things required in the top dog's make-up, and there are so few good dogs to breed from.

In dogs we may require fertility, size, speed, stamina, low requirement for water, short coat, intelligence, calmness, boldness, heading instinct, driving instinct, hold together, cover, square cover, the right sort of eye, bark (at the right time for the right reasons), the right type of bite, steadiness, width, break (breakaway and square), cast, mustering ability (scent, look, listen), perhaps backing ability, and more.

This is another reason why a policy of outcrossing is doomed to failure. You will never (not in one lifetime, anyway) breed top dogs by slow selection, as you can when breeding for only one or two traits in cattle for example. In cattle, if breeding for growth rate you can simply select the faster-growing cattle each generation and gradually breed a faster-growing line (although skilled inbreeding would get much better results). You could do this with one trait in sheep dogs, say 'strength', just keeping the strongest pup in each generation to breed from. However, while you are doing that you are neglecting other areas. So when you are breeding for maybe 20 or more various traits simultaneously (which you must if you want to breed good dogs), close breeding of some extent is essential.

Advantages of breeding dogs

Nevertheless, breeding sheep dogs does have certain advantages, and to be successful the breeder must make the most of these. One advantage is that dogs have a relatively short generation interval.

Another advantage is that dogs have litters. This means there is a much greater number of young to select from. Unfortunately many breeders negate this advantage by selling all of the pups except perhaps one, at six weeks of age, when they are too young to accurately assess their potential. In order to maximise our chances of breeding good dogs, the pups should all be retained until they are keen to work sheep (usually at four- or five months old), so that we can retain the best pup.

Remember that unless the pup you keep to breed the next generation from is as good as, or better than, its parents (or at least one of its parents), you're going backwards.

Another advantage of breeding dogs is that I believe the dog's genetic make-up is mostly there to see, for those with eyes to see. I believe that the traits we seek are mostly highly heritable (this is discussed in more detail in the next chapter). A cow's growth rate can be largely influenced by its feeding and nutrition, but I suggest that a dog's inherited abilities are mostly fixed, *provided you see it tested to its limits* so that you can ascertain what is natural and what is a result of training, though with experience you can generally tell.

You will not see dogs tested at sheep dog trials where the sheep have been quietened down beforehand, and where there are long time limits and poor design, or when they are working in yards. But if you are able to see the dog really tested on wild sheep and in challenging situations, and left to its own resources, its inheritance becomes obvious.

The breeder gains this advantage by watching their pups at work when they are young, out in the open without any influence from the handler, and selects them accordingly.

It has sometimes been said that 'It is not how they start that matters, but how they finish'. This may be true for pure trial work on easy courses with easy judges, but it is certainly not true for the breeder, nor for the handler looking for quality. Such dogs may work well when under control, *but when things get tricky they will let you down*. To the real breeder and stockman *how they start* is of paramount importance, because that tells you what their inheritance is, and how they will perform in difficult situations.

If you have to make excuses for a pup, 'Oh, he's just going through that puppy stage', or, 'I don't want to let him loose on sheep in case he gets into bad habits', then you know that the pup has shortcomings, and that it will *always have those shortcomings*.

How top dogs were bred in the first place

So, with all these difficulties facing the breeder, how is it that good dogs were ever bred in the first place? How did breeders of the past breed such outstanding dogs? I believe an understanding of this will help modern breeders to breed better dogs.

Necessity is the main answer. Necessity is the mother of good sheep dogs. Years ago stock were worked off a horse, and in the UK – where all our dogs originate – they were often worked by shepherds on foot, and in rugged country.

The early breeders wanted dogs that would muster out of sight or a great distance away in rough country, so that they didn't have to do too much walking or riding themselves; dogs that would force strongly, because they didn't have a motorbike to race around on honking the horn; dogs that would work one sheep, and cut and hold it off the others, so they could draft out in the paddock or so they could catch a ewe in lambing trouble while on foot; dogs that were controllable at long distances, and so forth.

So good dogs were essential. Thus the early breeders had a far better idea of what a good dog is than do modern breeders, who may only potter around with a few sheep in the backyard or on the trial ground, or do most of the work on a motorbike. Also the early breeders were breeding *solely* for practical usefulness.

But over and above these points, which are vital, I believe the other fundamental reason that they bred such good dogs was close breeding. Travel was much more limited in those days, as was communication. Families tended to stay within their own towns for generations. This meant that distinct bloodlines developed that may have been relatively pure for generations. Because of the lack of travel, inbreeding or line breeding would have been more the norm.

In recent times travel has increased exponentially, as has communication, so that now it is not uncommon for people to travel halfway across the country to join a bitch, mostly to a more or less unrelated dog.

In support of my view that close breeding played a significant role in the early breeding of working dogs, consider that most British breeds of livestock can be traced back to considerable inbreeding at their root. R.B. Kelley, author of *Sheepdogs – their breeding and management*, wrote another book entitled *Principles and methods of animal breeding* where he outlines the history of the formation of various breeds of animals. They all relied on some degree of close breeding, or 'in-and-in' breeding as it was called.

The following quote gives some idea of the widespread application of inbreeding, about the late 1700s:

> Prentice suggests that Bakewell [who bred the Leicester sheep and Longhorn cattle, and also horses and pigs] adopted his policy of mating like to like from the early procedure for breeding English gamecocks. It is also stated (Pawson 1957) that Hugo Meynell, first master of the Quorn hunt, only a few miles from Bakewell's farm, was breeding fox hounds in the 1760's on the same principles as Bakewell. Macdonald and Sinclair (1909) state that the first Tomkins who bred Herefords practised in-and-in breeding before Bakewell adopted the method. Thoroughbred horses also were 'under way' before Bakewell adopted the method. (Kelley 1960)

So you can see that inbreeding or line breeding was widely practised not only in cattle and sheep (as Kelley goes on to show also in the history of various other breeds) but also in foxhounds, in the late 1700s. It is highly likely, given these facts and the already mentioned limitations on travel and communication, that sheep dog breeders also used the method, both intentionally and not so intentionally.

The following is a quote from the 1920 book *Management and diseases of sheep in Australia*, first with regards to the selection of sheep dogs and then to their breeding:

> It is advisable to get a dog as much inbred as possible ...
> Should the breeder feel inclined to try breeding sheep dogs, let him select two of the best, preferably of the Australian Kelpie strain. They must be silent

workers, obedient and hardy. Then breed from this strain, always breeding-in as much as possible. All undesirables should be promptly culled out, those of bad colours and weedy appearance, also any which fail to come up to a high standard of working instinct.

Dogs have been bred successfully on these lines for upwards of twenty years. Any time an out-cross was tried the experiment proved a complete failure. (Cramsie *et al.* 1920)

The well-known breeders Mr King and John Quinn also utilised in-and-in breeding, as the following quote from Herb Morrice indicates:

Mr Quinn bred thousands of Kelpies, using only selected parents and commenced with one dog and two bitches. Mr King only provided very few and all related toward the King and McLeod Kennel. Biddy was the prominent one. She was mated with Coil and their progeny with Wallace. They were then well and truly inbred together to make up the Wilga Downs Kelpies and Barbs. These were very good dogs, which all speaks well for the good that must have been in the original importations. (Morrice, in Kelley 1942)

So I believe that there were two main reasons (in order of importance) why those sheep dog breeders of years ago managed to breed such amazing abilities into their dogs.

1   The first was necessity, and knowledge born out of that necessity and experience, and from observation and thought.
2   The second was close breeding brought about by the limitations on travel and communication, and by conscious plan.

A week-old litter of 'Campaspe' Kelpie pups – four red-and-tans; one black-and-tan; one fawn-and-tan; and two creamies

A knowledge of how those original breeders managed to put such remarkable abilities into their dogs, of which we see only a faint shadow today in most dogs, should help modern breeders to breed dogs worthy of their illustrious ancestors.

Present-day sheep dogs

It is a sad state of affairs when the quality of modern dogs is so poor. This applies to all breeds, although to some more than others.

However, the most important factor is not the breed per se (because the majority of dogs of all breeds are very poor), *but the individual dogs or lines within those breeds*. There is a great deal of mediocrity in all breeds, and only very few good dogs bred by the more knowledgeable. Each breed has a small number of good dogs, if you know where to look.

But whatever the breed, the same traits are necessary to breed a top dog. There is only one way a top dog can work, and any deviation from that is a fault or shortcoming. No matter what breed you choose or like (or even better if you ignore the artificial boundaries between breeds altogether), you must still select for the same traits.

Breed (or colour) isn't worth a drop in the ocean; what matters is the right natural ability.

11

Breeding methods

NOTE: In this chapter I have assumed that the reader has a basic knowledge of genetics.

There are, I believe, two 'secrets' required in order to breed the type of dogs that you are after.

The first is having a thorough understanding of a dog's inherited traits and being able to accurately assess dogs and pups, and from that knowledge then having a clear idea of the exact make-up of the dog that you are trying to breed. Without that clear, specific animal in mind you are doomed to failure before you even start. This was dealt with in Part 1.

The second secret is some degree of inbreeding or line breeding.

Cross faulting

'Cross fault outcrossing' is perhaps the most widely practised 'system' of breeding. The would-be breeder takes their bitch to a dog, most likely an outcross dog, which they think is strong in the points that their bitch is weak.

The result is mostly failure. *Always breed to what you want, and never to extremes.* For example, it is hopeless breeding a weak dog with eye and style to a strong, plain dog, hoping to get a strong dog with eye and style; or breeding a wide-working paddock type to a yard dog hoping for a 'utility dog'. It doesn't work. If you want utility dogs you must breed to utility dogs.

Many years ago, James Moore (Border Collie breeder and author of *The canine king*) had this to say:

> Never mate two extremes, for to mate similar and dissimilar together in order
> to obtain the mean must result in failure. (Moore 1929)

Kyle Onstott, a geneticist and dog breeder, wrote on the same point.

> Excellence in breeding cannot be achieved by the effort to blend two faulty
> extremes to arrive at a happy mean. If a fault is to be corrected in stock it must
> be balanced with excellence and not with its antithetic fault. The late Jack
> Holgate, successful breeder of so many varieties of dogs, whose Southboro
> prefix was one to conjure with, pointed that out in a discussion of Fox Terrier
> size some forty years ago. 'Don't put a big one to a little one to get the right
> size, because you will get big ones and little ones, not correct ones. Use the
> right size for one of the mates,' he explained in essence. (Onstott 1962)

So the most fundamental rule of breeding is to breed to what you want, and 'like-to-like' most of the time. However, in sheep dogs this is nearly impossible with an outcross, because there is so much variation. Almost by definition, if you are outcrossing you are breeding 'similar to dissimilar'.

You may breed 'like to like' for one or two traits with an outcross, but not for 20 or more.

Also there are so few good bloodlines, that if you want to breed from good dogs then you will have to stay more or less within one bloodline.

Onstott also said that:

> It is not the experienced and informed breeders who are forever experimenting
> with outcrossing strains. It is rather the novice and inexperienced who hope in
> one fell swoop to garner in one generation all of the virtues of both outcrossed
> strains and to eliminate their faults. (Onstott 1962)

You can breed the odd good dog by outcrossing, but *never consistently*, and it takes much greater knowledge (and/or luck).

Inbreeding

Inbreeding is basically close breeding. It has been defined in various ways, such as the mating together of two animals more closely related than the general population from which they come. But basically most would consider it to be the breeding together of brother and sister, or father and daughter, and so on.

Line breeding is simply a lesser degree of inbreeding, such as grandfather over granddaughter, or you may have the same dog appear a dozen or more times four or five or six or more generations back in the pedigree. So the difference between the terms inbreeding and line breeding in common usage is simply one of degree.

J. Lush, a well-known geneticist of the middle of last century, said that:

> more opportunities for breed progress are lost by not inbreeding when
> inbreeding would be advisable than are lost by too much inbreeding. (Lush
> 1945)

I believe that the exceptional ability of many top dogs has been wasted by a fear of close breeding.

Close breeding, as stated elsewhere in this book, is particularly needed in the breeding of sheep dogs because we are selecting for so many attributes at once.

However, its intelligent use is also valuable in the breeding of any animal, and I believe that selection as generally practised in livestock is far inferior to a sound policy of closer breeding to highly selected individuals. The very best animals are usually so far above the average that even if inbreeding depression of some level did occur (and it doesn't necessarily have to), the improved production would more than compensate for it.

Also, when outcrossing, your main tool is selection. This is why various 'heritability' factors have been calculated to show how much a given trait is likely to be affected by selection alone. These heritabilities calculate the percentage of the trait that is dependent on additive gene effects. A given trait might, for example, have a heritability of 40 per cent.

However, these heritabilities ignore the contribution of other gene effects, including 'epistasis' (an interaction between non-allelic genes), because they are assuming that the breeder isn't inbreeding. These epistatic gene effects are not very much affected by selection alone, not when you have so many traits to consider, but certainly are by inbreeding, and so inbreeding allows much greater improvements.

There are really two definitions of 'heritabilities' – one narrow, and one broad. The most common type is the narrow usage wherein the heritability percentage only takes into account the contribution that the additive gene effects make. The broader type of heritability factor also takes into account the non-additive gene effects.

I am certain that the heritabilities for most of the working dog's natural abilities are very high in the broad sense, but I have strong doubts about many of them being dependent on additive gene effects, although some do appear to be.

This is the main reason I believe close breeding is essential, hand in hand with the fact that with sheep dogs we must select for so many traits simultaneously, and that many of those traits are somewhat interconnected and interdependent (epistatic).

The effect of inbreeding

Inbreeding (of any amount) creates some degree of what is called 'homozygosis'. This simply means that on the same position (locus) on each of a pair of chromosomes (chromosomes act in pairs), the same form (allele) of each gene is carried.

It is important to remember that genes are transmitted to the progeny as part of the larger chromosome, and not as individual units for the most part (I say 'for the most part' because there is a phenomenon called 'crossing-over' which is estimated to result in roughly 10 per cent of gene transfer between chromosomes).

This means that 'strings' of genes are inherited together, not the genes individually.

I believe that this fact is not stressed nearly enough, because it has important ramifications. Bear this in mind when reading or thinking about genetics. (This is part of the explanation for 'throwbacks', and is part of the reason why I believe the emphasis placed on additive gene effects and heritabilities by most advisers is misplaced.) In the dog there are 39 pairs of these chromosomes, or 'strings of genes'.

To simplify things, imagine that a dog had only one pair of chromosomes (strings of genes), that is, two chromosomes. We will call them A and A1. Now imagine that we join this dog to a bitch with chromosomes that we will call B and B1. Each parent transmits only one of their pair of chromosomes to their offspring, so that the offspring ends up with a pair of chromosomes, one chromosome from the mother and one from the father.

The possible offspring in this case are shown in the following diagram, and have the following four possible chromosome combinations: A–B, A–B1, A1–B, A1–B1.

		Father	
		A	A1
Mother	B	A–B	A1–B
	B1	A–B1	A1–B1

Now if we take any one of these four pups, and inbreed it back to its father, so that we are joining the father A–A1 to the daughter A–B1, for example, the possibilities from this mating are:

		Father	
		A	A1
Daughter	A	A–A	A1–A
	B1	A–B1	A1–B1

So you can see that two from this mating (A–A and A1–A) have only chromosomes inherited from the father (A–A1), none from the grandmother (B–B1). That is, they are completely unrelated to their own grandmother (ignoring any maternal gene effects and so on).

Also we have one pup that is A–A, or in other words it has two identical chromosomes and is therefore completely homozygous, which means that every gene is the same on both chromosomes. This means it can only pass that chromosome onto all its progeny. Therefore it will breed consistently and may be fairly prepotent (have a stronger influence than normal on its offspring).

For example, in a relatively simple trait such as coat colour, there might be only two genes (one pair – one on each chromosome) that determine the main colour (although in practice the situation is more complex).

Say the father in the above example carried one gene for black-and-white on one chromosome, and one gene for red-and-white on the other chromosome. Because black-and-white is dominant and red-and-white is recessive, the father himself is black-and-white, but could transmit either to his progeny.

In the case above (inbreeding to the father) we ended up with a pup with two of the same chromosomes (A–A), therefore, depending on which colour was on which chromosome, the pup would carry either black-and-white on both chromosomes (in which case it would be black-and-white) or red-and-white on both chromosomes (in which case it would be red-and-white). This is what is meant by 'homozygosis' – the same genes on both of a pair of chromosomes, and this is what becomes more common with inbreeding.

This is why faults can pop up more often when inbreeding, because faults are mostly caused by recessive genes. This means we can get a red-and-white pup when neither parent was red-and-white (although this can still occur with an outcross but less often).

Of course the situation is more complicated than outlined above, because instead of each parent having two chromosomes (one pair), they have 39 pairs. So in practice we would never get a situation like the one above where the inbred pup was 100 per cent homozygous (although it is theoretically almost possible); rather we talk about averages and probabilities instead of absolute figures.

So when we calculate inbreeding coefficients (explained below in the section 'Degree of inbreeding – measurements'), for example the coefficient of pups from a father–daughter mating will be 25 per cent, this means that *on average* 25 per cent of the gene pairs in the pups will be homozygous, or identical forms on each corresponding chromosome.

Inbreeding tends to fix various traits in a line of dogs, and/or accentuate them. Consider the common belief that inbreeding causes small dogs, or excitable nervy dogs. If you inbreed to smallish dogs or excitable dogs you will breed a great proportion of such dogs, or perhaps breed even worse ones if you don't apply enough of the right selection pressure.

But if you inbreed to big, calm, bold dogs, and select in that direction, you will do the opposite.

Consistency and prepotency

From the discussion above you should be able to understand how inbreeding creates both consistency and prepotency in a line of dogs. By collecting the genes (or chromosomes) of the desired animal, and increasing their frequency in the line, and eliminating the genes (or chromosomes) of the undesired, you gain this consistency.

The prepotency comes from the homozygosis. For example, a dog that is homozygous for black-and-white can only pass that on to its progeny, whereas a dog that is heterozygous can pass on one of two possibilities. This means that inbred dogs generally breed more consistently than outcrossed dogs. However, how prepotent a dog is will also depend on whether it is homozygous for recessive genes, or for dominant.

If you outcross as a system, time after time, the results will swing wildly from one extreme to the other, much like a dog with no balance. There will be a lot of variation between the pups. But if you breed closely you give yourself a strong, consistent, prepotent foundation to work from, so that when you outcross the resultant progeny will be more influenced by your line than the outcross. If you then choose the best pup from the outcross, and inbreed or line breed it back into your line while selecting for the desired trait, you will be on your way to breeding what you want.

Inbreeding defects

The common argument against inbreeding is that it causes defects. It is true that some defects *may* occur, which is understandable when you consider that inbreeding increases homozygosis as explained above. Therefore, it is more likely to bring recessive genes together such as the red-and-white, which you are less likely to see in an outcross.

However, it is equally obvious that the inbreeding didn't cause the genetic fault, it simply brought it to light. Outcrossing simply hides these faults most of the time, although they can still occur.

Consistent outcrossing in fact results in an increase in the number of these undesirable genes in the population, because it hides them and allows them to be perpetuated unseen. Consistent inbreeding brings them to light, and so enables their reduction in the population by selection against them. Dutch geneticist A.L. Hagedoorn noted the following in his 1939 book *Animal breeding*:

> The bad effects of inbreeding, being due to heterozygosity of the stock, are mainly due to cross-breeding in earlier generations. This is one of the paradoxes we meet in this subject. In-breeding produces variability and the production of undesirable individuals, but the only remedy is not out-crossing, which lays the foundation for much more impurity, but more and long-continued in-breeding... (Hagedoorn 1939)

The best way to know if a dog carries defective genes is by inbreeding to it. So if you find a good enough dog, you join him to the best of his own daughters, and consider the result. You might find that you get no problems at all, that the pups are fit and healthy and fertile, in which case you can continue inbreeding with relative impunity. Combine that with wise selection and you will create a very strong bloodline, largely free from inherited defects.

However, if you do find that some defect occurs, such as deafness or eye problems or something of that nature, you would avoid the dog in question, and try another one. You would not perpetuate the defective genes.

But if you were outcrossing you wouldn't realise that the dog carried those bad genes, and so you would continue to use him and spread the genes around.

Inbreeding depression

Inbreeding depression is the generalised decline in health and vitality that can accompany inbreeding, such as a reduction in fertility, disease resistance, growth rates, and so on. It often discussed when inbreeding is the subject. It can be a problem, however it is not a foregone conclusion that it *must* occur as many people believe. It does occur with *random* inbreeding when selection against it is minimal or non-existent.

Conversely, inbreeding plus sound selection can produce an increase in the vigour of the population.

Regarding a famous experiment in inbreeding, Onstott said:

> An experiment in the prolonged inbreeding of white rats has been carried on
> by Miss King, a worker at the Wistar Institute. For twenty generations she bred
> together full brothers and sisters, which is recognized as the most intensive
> form of inbreeding, selecting her breeding stock constantly for its vigour and
> stamina. At the end of twenty generations of intensive inbreeding, Miss King
> had a race of rats of greater than average size, greater fecundity [fertility], and
> greater longevity than the stock with which the experiment had begun.
> Inbreeding plus selection resulted in the improvement of the strain. Without
> selection, this would, of course, have been impossible. (Onstott 1962)

It should be noted that brother/sister breeding such as that above is more amenable
to selection than father/daughter breeding, because you choose two completely new
animals each generation instead of breeding back to the same animal. It is a different
joining, used with a different intention in mind.

Inbreeding, when used wisely and combined with careful selection, is extremely
valuable. However, it must be used wisely, with an understanding of the risks, *but these
are not as great, or as common, as many people think.*

In fact, more distant line breeding can be more problematic than close inbreeding,
because you may not realise that a dog has a fault until you begin doubling up on it a
number of generations later, *and by then it may have sired numerous animals.* On the
other hand, if you inbreed closely to it you know fairly quickly what it does, or does not,
carry in the way of defects.

You must place health, stamina and fertility high on your list of priorities when
breeding. If you ignore them they will cause you (and others) trouble.

Degree of inbreeding – measurements

As mentioned earlier, it is possible to calculate how inbred a dog is. This will give us a
percentage figure relating to how many gene pairs are homozygous, or pure breeding.

Inbreeding only occurs if the same animal occurs on both sides of the pedigree.

Thus in some cases you might join two highly inbred animals of two separate
bloodlines (an outcross), and you will get zero inbreeding in the pups even though both
parents are themselves highly inbred. Therefore in any inbred line of dogs you can
immediately bring the inbreeding to zero by an outcross.

The first step in calculating a dog's level of inbreeding is to examine the pedigree and
determine which dogs occur on both the father's and mother's sides. Consider the
following pedigree of Moorlands Georgie:

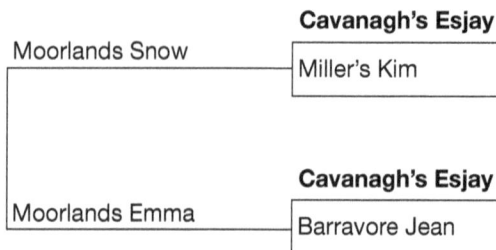

		Cavanagh's Esjay
Moorlands Snow		Miller's Kim
		Cavanagh's Esjay
Moorlands Emma		Barravore Jean

You can see that Cavanagh's Esjay appears on both sides of the pedigree. Therefore there is inbreeding to him. We can write the inbreeding as 2:2 to Cavanagh's Esjay, because he is two generations from Georgie on her father's side, and two on her mother's (the colon is used to separate the occurrences of the ancestor on the father's side from those on the mother's).

We can now use geneticist Malcolm B. Willis's method to calculate the inbreeding coefficient. The complete formula is:

$$F_x = \sum \left[\left(\frac{1}{2} \right)^{n1 + n2 - 1} (1 + F_A) \right]$$

To decipher this, F_x is the inbreeding coefficient that we are trying to calculate. $\sum$ means 'the sum of'. $n1$ and $n2$ are the generations from the dog in question back to the dog being inbred to. F_A is the inbreeding coefficient of the dog being inbred to, for example we may be inbreeding to a dog which is itself already inbred.

The total formula tells us that in order to calculate the inbreeding coefficient F_x of a dog, we need to add all of the individual pieces of inbreeding together ($\sum$ = 'the sum of').

So to take our above example, we have inbreeding to Esjay of 2:2, because he is two generations back from Georgie on both sides. This 2:2 gives us our $n1$ and $n2$ figures, which we add together and subtract one ($n1 + n2 - 1$). So $2 + 2 - 1 = 3$.

Jack Hiscock with Cavanagh's Esjay. (Photo courtesy Joan Hiscock.)

We now raise $\frac{1}{2}$ to the power of 3, which gives us 0.125, or 12.5 per cent. Cavanagh's Esjay is not himself particularly inbred, so we can ignore the $1 + F_A$ component of the formula. If he was inbred we would multiply this 0.125 by 1 plus his inbreeding coefficient ($1 + F_A$).

Because this is the only inbreeding in the pedigree that we are worrying about, we don't need to add any other inbreeding to it, so we can ignore the 'sum of' portion of the formula. Thus Moorlands Georgie was inbred 12.5 per cent to Cavanagh's Esjay, which is to say that she had roughly 12.5 per cent of her genes homozygous. However, in theory she could have had as little as almost 0 per cent inbreeding (homozygosis) or as much as nearly 100 per cent, but chances are that she had somewhere near 12.5 per cent. It is a statistical probability only, not an absolute figure.

However, things can get much more complicated. If we extend this pedigree back another few generations, further inbreeding (or line breeding) becomes evident:

	Cavanagh's Esjay	Cavanagh's Craig	**Miller's Pete 2nd**	*Yulong Russ*	Sinclair's Butch
				Moorlands Tracey	
		Cavanagh's Teena			
Moorlands Snow					
		Yulong Russ			
	Miller's Kim				
		Moorlands Tracey	Sinclair's Butch		
MOORLANDS GEORGIE					
	Cavanagh's Esjay	Cavanagh's Craig	**Miller's Pete 2nd**	*Yulong Russ*	Sinclair's Butch
				Moorlands Tracey	
		Cavanagh's Teena			
Moorlands Emma					
		Miller's Pete 2nd	*Yulong Russ*	Sinclair's Butch	
			Moorlands Tracey		
	Barravore Jean				
		Timaru Belle			
			Minnieburn Dell	*Yulong Russ*	

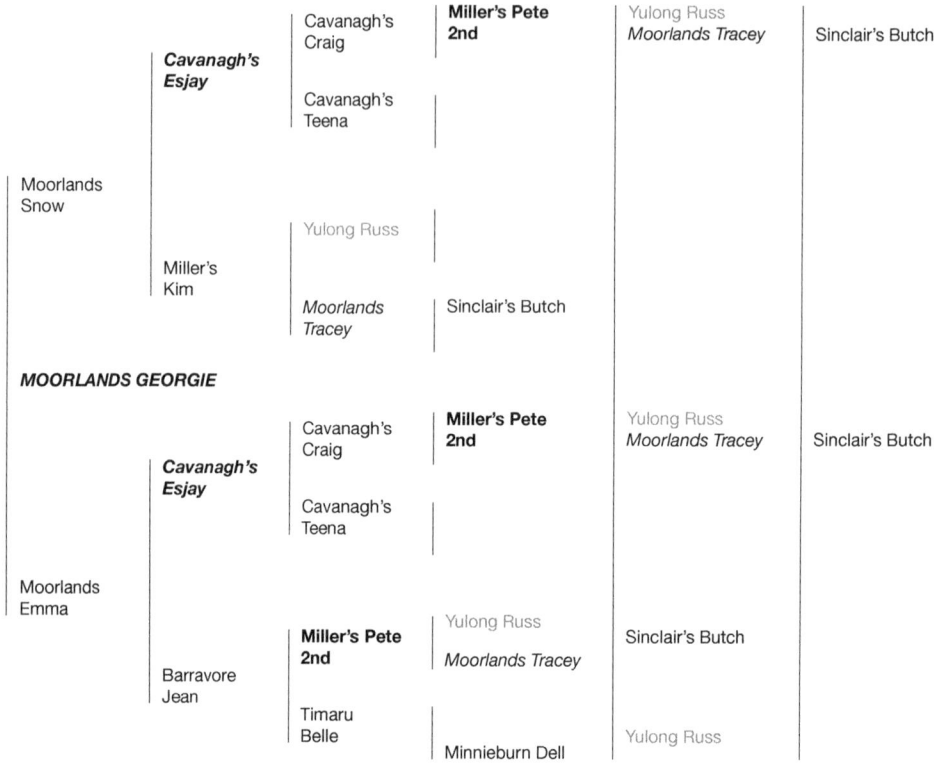

You can see that we also have inbreeding to Yulong Russ, Miller's Pete 2nd, and Moorlands Tracey. Sinclair's Butch appears a number of times, but always through Moorlands Tracey, so his contribution will be calculated via her, and can therefore be ignored.

It becomes complicated because we have to take into account the inbreeding already calculated. So we ignore the inbreeding to Miller's Pete 2nd for example, if it comes through Esjay on both sides, because we have already taken that into account.

So we start again by writing out the generations from Georgie to each dog that is being inbred to:

Cavanagh's Esjay 2:2
Miller's Pete 2nd 4:4.3
Moorlands Tracey 5.3:5.4
Yulong Russ 5.3:5.4.5

First, let us consider Miller's Pete 2nd. Two lines come through Esjay, so we write it like this: 4:4.3, underlining or putting a line through the lines on each side that come through the same dog (the decimal point is used to separate multiple occurrences of the ancestor being inbred to, either on the father's or mother's side of the pedigree). Now when calculating the formula we only worry about those lines on each side of the colon that don't *both* have a line under them, that is, where both lines don't come through Esjay, who we have already calculated. So we calculate the inbreeding to Pete as follows:

$\frac{1}{2}$ to the power of $\underline{4}$ + 3 − 1, because the two lines of inheritance came through different dogs. We don't calculate the $\underline{4}$ + $\underline{4}$ because both of these lines come through Esjay.

So the inbreeding to Pete (ignoring the inbreeding to Pete which has occurred through Esjay) is $\frac{1}{2}$ to the power of 6, which is 0.0156, or 1.56 per cent. As you can see it is very small, because it is getting a fair way back in the pedigree.

The case is similar with Moorlands Tracey 5.3:5.4. Three of these lines come through Miller's Pete 2nd, who we have just calculated. Thus $\underline{5}$.3:$\underline{5}$.4. So we only calculate those numbers on either side of the colon that aren't both underlined. We calculate each line of inbreeding separately, and then add them all together; this is where Σ comes into play. Thus only 3:$\underline{5}$ and 3:$\underline{4}$. So 3 + 5 − 1 = 7, which is 0.78 per cent. 3 + 4 − 1 = 6 which is 1.56 per cent. We then add these together to get the total inbreeding to Moorlands Tracey (ignoring her effect through Miller's Pete 2nd) and we get 2.34 per cent, again fairly minimal.

With Yulong Russ we have 5.3:5.4.5. Three lines come through Miller's Pete 2nd, but two come through different dogs (Miller's Kim and Minnieburn Dell). So $\underline{5}$.3:$\underline{5}$.4.5. Therefore we calculate each one. So $\frac{1}{2}$ to the power of 9 ($\underline{5}$ + 5 − 1) which is 0.19 per cent; 3 + $\underline{5}$ − 1 = 7 which is 0.78 per cent; 3 + $\underline{4}$ − 1 = 6 which is 1.56 per cent; and 3 + 5 − 1 = 7 which is 0.78 per cent. If we add all of these together we get 3.31 per cent.

So we have inbreeding to each dog as follows:

Cavanagh's Esjay 12.5%
Miller's Pete 2nd 1.56%
Moorlands Tracey 2.34%
Yulong Russ is 3.31%

If we add all of these up we arrive at an approximate total figure for Moorlands Georgie's inbreeding coefficient of 19.71 per cent, which tells us the percentage of her gene pairs that are likely to be homozygous.

Line breeding

Line breeding is simply less intensive inbreeding. Strong inbreeding is valuable when you have an exceptional dog to inbreed to, which has no bad faults. But this is rarely the case.

Say for example you have a top dog but he is on the small side. Close inbreeding is likely to be unsatisfactory because you will end up with small dogs. If you had enough good bitches to inbreed back to him, you may be able to apply enough selection pressure and get enough dogs inbred to him that are a better size.

The other option is to breed a bit wider. The other blood should be strong in the area that he is weak in, so you would choose good tall dogs, as good as possible in every other area also, and particularly strong in any other area he might be slightly deficient in.

If you then get what you require in the line bred litter, that is, a top dog with more size, you might then inbreed more closely to that.

When breeding you must be flexible, as the following quote suggests:

> Some breeders outcross their dogs (or think they outcross them) through fear of the degeneracy of type that they believe results from inbreeding. There are others who line breed (or think they do) in an effort to obtain the good results of inbreeding without assuming its alleged, and sometimes real, hazards. These men are like the Irishman who cuts off the puppy's tail a joint each day so that it will not hurt him so much as to cut it all off at once. There are yet others who, having been told that good dogs are produced by inbreeding, accept inbreeding uncritically as a matter of policy. They inbreed (and know that they inbreed) merely for inbreeding's sake. A fourth class inbreed one generation to obtain their type and outcross the following two generations to restore to their stock whatever stamina may have been lost by the inbreeding process. All of these kinds of breeders sometimes produce good dogs, despite that all of the policies and reasons for them are ill-founded.
>
> There are, of course, a few breeders who know exactly what they are doing. They inbreed when inbreeding is indicated; they line breed with a purpose; and they outcross to obtain certain characteristics which they desire to annex to their stock. (Onstott 1962)

Line breeding to a distant ancestor

Often people line breed to some famous dog back in the pedigree, but this can be fraught with two main hazards. The first is that you are line breeding to it *through other dogs*. Now if each of these other dogs is a different dog, that is all right. But mostly the same descendant of the desired animal pops up time and time again, and so instead of line breeding to the dog you are after you are in fact line breeding to one or more of its descendants.

For example, I like as much Yulong Russ blood in my dogs as I can, but often this blood is coming through only two or three other dogs, such as Millers Pete 2nd, for example. I don't mind this, because I don't mind line breeding to Pete, but with other dogs in other situations it may not be desirable.

The other big problem arises if we line breed to a famous dog that we ourselves never saw work (or only a limited amount) and therefore we may not have a clear idea of its abilities. Hence we can be line breeding to one type of animal and selecting for a different type of animal, and so be fighting against ourselves.

Also it is important not to place too much emphasis on animals far back in a pedigree, because there has been a lot of selection – and probably deterioration – between them and the litter in question. The dogs close-up are *vastly* more important.

Close breeding and combinations of traits

As explained in Part 1, there is considerable interplay between various traits, and also between the traits and the dog's temperament, and a huge variation in the actual make-up of the dog in any area.

This is one reason why breeding often fails, because the breeder doesn't understand this complex interplay, for example the fact that strong eye and style can be directly associated with weakness or tension, or that a keen heading dog is often made that way by an excitable nature.

As I have said before, a good example is when a breeder attempts to join a wide-working, strong-eyed, stylish but weak bitch, to a strong, calm, plain dog in the hope of getting something with the best points of both. It cannot occur because it is the association between the various traits that causes the dog's final appearance, and therefore the strong pups will be plain, and the weak pups will have the eye and the style.

Also every dog or bloodline has various traits in certain combinations, for example a good dog may have calmness and strength combined with 'breakaway break' and real cover. Each bloodline differs in its combination of traits. Some will have a combination of eye and bark, others will have a combination of eye and strength, others only of eye and weakness.

If you breed closely you maintain these combinations, and so will tend to reproduce the dogs you like much more consistently, instead of destroying complex associations which may have taken generations to assemble. You can create new combinations of traits, but it takes time to fix them.

Another good example is to assume that you decide to breed for casting ability. You may join your bitch, which you consider to be a good caster, to an unrelated dog that you also consider to be a good casting dog. But it is quite possible that the two animals cast 'well' for two different reasons. Your bitch might cast because it is weak and breaks very wide (counterfeit cast), and you have only tested it casting in small areas. The strong dog might cast well because it has a true natural cast, and scope.

You join the two together, and very few of the pups have natural cast, if any. Most don't cast like the bitch, because most are not quite as weak as she is, and the associations which made the wide break and so on have been broken up. And most don't cast like the dog for the same reason, because his natural casting ability has been broken up by the lack of it in the bitch.

So you have joined two supposedly good casting dogs and yet the litter isn't much good at casting.

On the other hand, if you breed the bitch to her own father, who has the same type of counterfeit cast as she does, you would get a litter of pups many of which would 'cast' like the bitch, even though it isn't real cast. So, close breeding is a way of producing the type of dog you see in front of you, even if you fail to understand why it works like it does.

If you wanted real cast, you might join your bitch to the real casting dog, and even though none of the pups would be any good at casting you should retain the best bitch pup, and join it back to its real casting sire. In this litter the associations that made up the real casting ability would *start* to come back together, and you would *begin* to see some of the real casting ability, but it may take another two or more generations to mostly remove the undesirable influence of the poor-casting bitch you started with.

There are two points to take from this example. One is that it wasn't the dog's fault that the first litter was no good at casting. Some people would say that the dog was a good caster, but never bred on, but this isn't the case. It wasn't the dog's fault, it was the breeder's.

The second point is that breeding is a long-term process, and you will not breed the perfect dog in one generation with an outcross. You must be patient and work towards your goal one step at a time.

The outcross in a close breeding program

The main use of an outcross is as part of a planned program of inbreeding or line breeding. You build your closely bred strain of dogs around the best dog or dogs of the best blood you can find, and breed somewhat closely to fix the characteristics you like, while applying strong selection pressure against any that do not measure up.

Then you might decide that your bloodline, although consistent and fairly high quality, lacks in some particular area. Perhaps you want to introduce some natural barking ability but without sacrificing the qualities you have already fixed.

The best bet would be if you could find a good dog fairly closely related that had barking ability. But let's assume that you cannot find such a dog, and that the only way to introduce barking ability is by a wide outcross, or by crossbreeding to another breed.

You must look for a dog that is high in the quality you want, but that is also of as high a quality as possible in every other regard, because faults are very difficult and time consuming to breed out. In fact, I am sure that it takes at least four generations before the influence of one poor animal can be mostly negated.

Hagedoorn, whom I quoted earlier, aptly described this situation:

> When it is unavoidable to employ an out-cross, we always get very much more than we need in the way of genes foreign to our breed or strain. We may obtain the factor we want, but in addition we also obtain a number of factors we certainly do not want, and after each out-cross the slow process of weeding out undesirable individuals will have to start all over again. (Hagedoorn 1939)

Therefore you must be very careful when outcrossing that the animal is of high quality. So you look around until you find a dog that is a natural barking dog, and also fairly good in most other respects, with no outstanding faults.

You must of course be careful that the ability you seek in the outcross dog is not due to some fault. For example, it would be no good breeding to an outcross that barked out of weakness or out of excitability (as most do), because then when you selected for calm dogs and strong dogs in the generations after the outcross you would lose the bark.

So you then join a good, closely bred bitch to this outcross dog, retain all of the pups, and select the best pups with the barking ability. You might then try breeding brother to sister from this litter, to fix the new combination of traits and to put the genetic associations back together which have probably been split up, if you consider the outcross dog to be of high enough general quality.

You could also try inbreeding back into your own line and selecting for the barking ability. However, if you inbreed too strongly back into your own line you might lose the bark; some degree of close breeding will be necessary, but not too much – only enough to restore all of your previously fixed traits while retaining the bark. So you will probably line breed to some extent.

When you have a dog with the barking ability and all of the previous ability, you then begin inbreeding or line breeding to that.

This is how you bring in a trait that may be lacking. It may be anything – bark, size, stamina, whatever, but it doesn't happen overnight. But very few people have a long-term plan for their breeding – they only think as far ahead as the next litter. You must look further down the track, and know *exactly* what you want to achieve.

The strength (or weakness) of a kennel is in its bitches

The bitches are most often the weak link in any breeding operation, and in most pedigrees. There are two reasons for this: first, the breeder must own the bitch (but can take it to be joined to most any dog he likes); and second, more bitches are needed than dogs, because one good dog can sire litters from many different bitches.

Thus there is usually much less 'selection pressure' applied to the female side of things, and more poor bitches are bred from than poor dogs. The most common reason why pups turn out disappointingly is because the bitch is poor.

You are unlikely to breed a good pup from a poor bitch, and so most often you are better off not breeding from a bitch at all unless it is of high quality. Some people breed from a bitch because it is 'well-bred', even though it is no good in itself, which is unwise. One television ad summed it up very well – 'It's the fish John West rejects that makes John West the best' – it is the same with dog breeding. The main reason that there is so much mediocrity about is because such inferior animals are bred from.

One story I have heard is of a handler who walked off a trial ground after a poor run, and commented on what a fantastic pedigree his dog had. An old-timer sitting on the sidelines replied, 'Well, next time bring the pedigree and leave the dog at home'.

The pedigree is only as good as the dog it represents.

If you do breed from second-rate bitches (perhaps because that is all you can obtain) then you must be willing to spend a number of generations in improving their quality. For example, you might breed the original poor bitch to a good dog, and with luck get a bitch pup that is an improvement, even though it is still not up to standard. Then join *that* to a good dog of the same line and type, and so on, until you eventually end up with a good bitch. But it will take years of work.

The bitches are the strength, or most often the weakness, of any kennel. A good bitch is a rare and valuable thing.

Summary

In summary, the basic method for breeding high quality sheep dogs is first to have a very clear idea of what you want. Remember that weak dogs, soft dogs, or excitable dogs are fairly useless, and that weakness and eye can give an impression of ability that isn't really there.

Don't be overawed by a flashy bit of style either, unless it is founded on real substance, and only used when necessary. And remember not to neglect speed and movement and stamina.

As a general rule you line breed to the best dog of the best line available, inbreed to an exceptional dog when it occurs, and outcross very rarely only with a specific purpose in mind.

You take a slightly different approach if you are trying to breed 'the perfect dog' rather than just wanting to breed a consistent line of handy dogs for general sale, but the differences aren't great.

Remember that some animals may possess too many hidden undesirable recessives, and inbreeding to them may cause problems. You will have to avoid them. For example, one dog I know of, when inbred to, produced a percentage of deaf pups, and poor moving pups. Another line has fertility problems. Other dogs will not have these problems. The only way to know is to try.

Without a clear understanding of exactly what you want in a dog you can never achieve it. Breeding is not a short-term process, it is a long-term one – you won't get perfection in one generation. Be patient.

The most important point is selecting the right animals to breed from, and avoiding the wrong ones. Everything else is secondary to that.

A Kelpie

Finally! A Kelpie
Like the dogs of old.

When break and hold and cast
were the rule
And 'yard dogs' were disdained.

When 'splitters' and 'chasers'
lived short
And stockmen knew their trade.

Finally! A Kelpie,
With speed and stamina
Amidst the morass
of the soft and sluggish.
A genius plying his trade.

With
skill and tact and touch.
Hard of body and sharp of mind,
Lean and tough and wiry.

As the others
yap and split and drool.
And their owners still proclaim,
"He's got cast and cover,
work any stock, any job, anywhere;
a champion indeed."
Kelpie in name only.

Ah! A *real* Kelpie,
Finally,
Like the dogs of old.

Part 3

PRINCIPLES OF TRAINING AND CONTROL

12

How dogs learn

In this chapter I would like to delve into the principles behind dog training, and how these can be applied in a practical sense to train a dog in the quickest, easiest, and most effective manner. The harder the dog is to handle the more important it is to follow these principles, and the greater the consequences are for breaking them.

Simply put, a dog does something for one of three reasons:

- to gain something it wants
- to avoid something it doesn't want, or
- out of habit.

The most effective training utilises all three factors, and creates an obvious distinction between the first two points.

If you desire to change a dog's habit you must work in these three areas. You do this either by making something else more desirable so that the dog wants to do *that* more than the bad habit, or by making the bad habit less desirable, and/or by creating an incompatible habit to take the other habit's place.

So to teach a dog to 'go back' and stay on the other side of the mob, you make it undesirable for the dog to be on your side by growling at it and chasing it every time it comes around your side. Once it is around the far side you leave it alone and it gets to work the sheep, so it likes that. And then you repeat this process until it becomes a habit.

What you must realise is that there are many factors that come into play, and that influence these three areas. It is the handler's 'art' to be able to read and understand and use these to their advantage – you have to be able to understand the dog's motivations. In a controlled situation, such as when teaching a dog to sit, it is fairly simple, but out in the real world of stock work the dog's motivations become much more complex, with many more variables.

The simple rule is to make what we want benefit the dog, and to make what we do not want disadvantage the dog.

When teaching something, the clearer we can make the distinction between what we want, and what we do not want, the quicker the dog will learn.

And aside from these things we must also utilise habit strongly, by repetition, to fix the desired response.

Associations

Dogs, and animals in general, and people, create associations in their minds all the time. For example, in people certain sounds or smells can bring memories to mind years after the actual event took place. The sound of a brass band may evoke memories of country shows, and so on.

Another important example is in the recognition of faces. Often if we have only met someone in a certain setting, and later see them in an unrelated setting, we recognise their face but we can't think who they are. This is a case of our mind having associated their face with the environment in which we saw them, and we have trouble placing them in a different environment.

Dogs are no exception, and they create associations all the time to all manner of things. To achieve good results the handler must be aware of this, because all learning is about creating associations.

For example, when teaching 'sit down' you may think that you are teaching the vocal command, but the dog will also associate the tone of your voice, your raised hand, your position standing close by, the lead rope, and the location.

And, in fact, some of these other factors will assume much more importance in the dog's mind than the vocal command, in particular the lead, your position standing over the dog, and your raised hand. You will have made a *start* in teaching the vocal command, but that is all. So if you then try to get the dog to 'sit down' using a different tone of voice, or without being close by, or in a different location, and so on, it has no idea what you are talking about, just as when you don't recognise a familiar face in a different setting.

Over time the dog gradually sorts out these factors, and narrows them down to those that occur at all times, and disregards those that don't. For example, as you give it lessons in different places it will start to disregard location as one of the cues. And as you use various tones of voice it will start to disregard tone as a cue, and learn to obey the words more or less regardless of the way they are spoken.

This is why it is important, particularly in gaining control at a distance, to make sure that the dog understands and obeys the vocal or whistled command on its own, without any influence from your proximity or body language or arm signals or anything else.

The dog will also learn to associate a growled command with correction, and a friendly voice with praise. This is not something that they know instinctively, although the bitch can teach pups to respect a growl when they are young, but you can see from watching pups that they don't know this instinctively. They learn to associate their mother's growl with being bitten, often fairly hard.

An interesting aside (for those who don't believe in punishment or correction) is that animals don't make much use of positive reinforcement to train their young. They use correction and punishment. The reward comes from being left alone. Correction and punishment are important factors in how animals learn in the real world, in their natural environment, as is reward. Some people claim that 'positive reinforcement' methods are 'natural', and that correction or punishment are not, but you can make up your own mind about that. The good handler makes use of all three (reward, correction and punishment), but the less punishment the better.

Creating associations

By way of demonstration, you can fairly easily create an association between a word and an action. If you were to go into a darkened room with someone and watch their eyes, and then switch the light on, their pupils would contract. If you say 'contract' (or any word you like) at the same time, and repeat this exercise over and over, you could then say 'contract' without switching the light on, and their pupils would contract. You would have created an association, sometimes called a conditioned response.

When training a dog you can do the same thing. If the pup is coming to you, say 'here'. You aren't directly training it, but you might create an association.

What you must realise, however, *is that you will not get control in this way*. In the case of the contracting pupil trick you are working with a subconscious response; that is, the person has no conscious control of their pupil size. With something like 'here' it is a different matter, because the dog can choose to obey or not, and there are many conflicting motivations which may pull it in other directions. So just creating such an association will not result in obedience to the command.

This association phenomenon can be very powerful, and you can teach the dog the exact opposite to what you think you are. For example, if you tell it to 'steady' every time it starts to go too fast, without enforcing that it slows down, you will quickly teach it that 'steady' means 'go fast'.

The same often happens with dogs that chop in as they run around sheep. The handler yells 'get out' as the dog begins to chop in, and so in the dog's mind 'get out' comes to mean 'chop in'. Another common example of this is in 'forcing'. Telling a weak dog to 'push up', when it is frightened and won't go any further, only causes it to associate 'push up' with exactly *not* pushing up.

A similar thing often happens when people try to get untrained young dogs to walk into sheep that are backed up against a fence. If the dog does come in and the sheep break past it, 'push up' can come to mean 'get tense and ready because these sheep are going to break', and so instead of walking in calmly the dog gets tense. Walking in to a couple of sheep backed up against a fence may be a good demonstration with a strong, well-trained, tight-holding dog, and a good test for an experienced dog, but not much good for a pup (particularly a weak pup).

Timing of commands to create associations

If you are trying to create an association in the dog's mind, you shouldn't give the command *after* it has done something, because it will simply associate the command with doing what it is doing at that moment. You should give the command *just before* it is about to do something. This may sound obvious, but most people give the command too late.

One time you might want to create an association is when teaching a dog to break out on the cast. If it is going to break itself out around a dam or something, tell it to 'get out' just before it breaks out, as early as you can when you see that that it is going to break out. The act you want is the actual break out, so it is no good continuing to repeat 'get out' after the dog has broken out and is then going around the dam. There is only time to give one command, and it must be given at the right instant, fractionally before the dog kicks out.

If you are teaching the dog to 'heel', you should give the command just before you make it come back into the right position behind you. Don't keep repeating 'heel' when it is walking along behind you nicely, *because it doesn't change anything in response to the command*, and so it learns to ignore it. Only give the command when the dog is to return to the heel position after it has moved away, just before you make it do so.

Give the command, then give the dog a chance to respond, then enforce the command.

Some people advocate using a command as praise for obeying that command itself. So they call the pup 'here' and then keep repeating 'here' when it arrives, as praise. Instead you should use a separate praise expression such as 'good dog'. Otherwise the pup is learning to ignore 'here' when you keep repeating it for no reason.

The same applies when teaching the directional commands. Don't keep repeating 'round' (to move clockwise around the sheep) while the dog is going round, because it doesn't alter its behaviour in response. Only give the command if the dog slows down or hesitates, in order to speed it up again. Otherwise it learns to ignore the command because it doesn't have to take any action in response to it. Also, if you keep repeating the directional command while the dog is going in the required direction, you will develop a situation where you *have* to keep giving the directional command to keep it moving.

So give the command once, and don't repeat it unless the dog slackens off or breaks the command.

The same applies when teaching 'walk in' to force. Don't keep repeating 'walk in' while the dog is already walking in. Only give the command if it slows down or hesitates, when you are in a position (such as with a long rope on the dog) to make sure that it responds to the command. If the dog is already walking in and you give the command, it should respond by speeding up.

Habit – the key ingredient

Habit and repetition are powerful tools in the dog handler's tool kit. Putting a command to an action once will not create the association in the dog's mind; only when it is repeated will the dog start to realise that the command isn't just a chance happening, but that it is associated with one specific action. Some dogs require a great deal of repetition, while others require hardly any.

Boho Misty, a yellow-and-white Border Collie bitch that I had, was exceptional in this regard. You only had to show her something once and she knew it. I have an eight-month-old pup now, Glendaloch Craig, who is similar. With most other dogs you have to repeat it over and over again.

Many UK Border Collies, bred for their trainability and learning ability, learn like lightning. They form habits very rapidly. Australian Border Collies and Kelpies are generally inferior in this regard.

So remember to use repetition whenever you are having trouble teaching a dog something. You may not think that it is ever going to get it, but sooner or later the penny will drop and away you will go. Have perseverance.

Positive reinforcement – its uses and limitations

Positive reinforcement means giving the dog something it likes (a reward) when it does what we want, such as giving it a biscuit for obeying 'sit down'. It can also include allowing the dog to work sheep as a reward, or many other things.

For example, if we make the dog 'sit down' before casting it out, and then cast it out with 'go back', the 'go back' is positive reinforcement for remaining lying down. But if we say 'go back' just as the dog is getting up itself without having been told, we will reward it for getting up itself without being told, and it will do so more often in the future.

If you tell a dog to 'stop there' (standing), and then only give the next command when it is standing still and relaxed, you will reward it for standing still and relaxed. But if you give the next command while it is jiggling around and not really standing still, then you will reward the fidgeting behaviour. Timing is crucial.

Another example is when picking a pup up. If the pup starts to struggle and wriggle people often put it down. But this has simply rewarded that behaviour. What you should do is wait until it relaxes, and then put it down.

Just about everything you do has an effect in rewarding or correcting the pup's (or dog's) behaviour, and therefore in shaping the way it behaves.

Some people nowadays go to great lengths to avoid correction, with the fashionable positive reinforcement training systems. However, as Alfie Kohn showed in his book *Punished by rewards*, every reward system also naturally contains correction, whenever that reward is withheld.

So if you have given the dog a biscuit every time it obeys 'sit down', then if sometimes you don't give it a biscuit, it has effectively been corrected. Imagine yourself in the dog's position. If someone said to you, 'Go and dig a hole, and if I like the hole I will pay you 10 dollars.' So you go and dig the hole, and he likes it, and he pays you 10 dollars.

He says the same thing again, and you dig a second hole. This time he says, 'No, I don't like that hole, I'm not paying you for this one.' You have been corrected, or punished, by the withholding of the reward. This demonstrates how positive reinforcement systems contain correction. The 'positive reinforcement people' fail to realise this (as well as many other factors discussed below).

Positive reinforcement systems contain correction, and correction systems contain reward. In fact *all* methods of effective training contain elements of reward and correction.

In sheep dog training we make use of positive reinforcement all the time, to shape the dog's behaviour. However, by itself this will never give you CONTROL.

Positive reinforcement fails when the dog's motivation to do something else is very strong, because the dog is rewarded for disobeying you. If you offer it a biscuit for obeying 'here' when it is eating a leg of mutton, you won't have much control, and it will be rewarded for ignoring you by continuing to eat the mutton. The same applies with stock work. If you tell a dog to 'sit down' on the cast, and it ignores you, it is rewarded because it gets to work the stock.

So you must make use of other techniques in order to get control. You need to use correction, or very occasionally punishment, just as animals do in the wild. If a pup is eating the bitch's food, or trying to get a drink from her when it isn't feed time, or chewing her tail, it soon learns to obey a growled warning.

However, some people don't make enough use of positive reinforcement, by always being down on a dog so that it never enjoys its work. It is important to allow the dog to be rewarded for its obedience by giving it the freedom to work, within certain boundaries. Gradually work towards full control and obedience, bit by bit, not all at once, otherwise you remove any positives for the dog and so it will become sour (see 'End of lesson rewards' later this chapter). How fast it becomes sour will depend on its temperament and keenness.

Commands as rewards and corrections

It is important to realise that just as commands can act as rewards, so too can they act as corrections, and their deliberate use as such can effectively teach a dog.

For example, any command that impedes the dog working the sheep tends to act as a correction, while any command that increases its contact or freedom with the sheep tends to act as a reward. Thus 'sit down' can act as a correction, but almost anything else that lets the dog move off and work the sheep acts as a reward.

Commands only tend to act as correction when they are used repeatedly in the same situation, so that the dog realises that they are associated with that situation. So if we tell a dog to 'steady' and it doesn't slow down, we can then tell it to 'sit down' as a correction. This is fairly effective *when repeated over time.*

But 'sit down' is not very effective as a correction for breaking a previous 'sit down' command. This is because the dog's situation has improved by ignoring the first command, because it has moved a few metres towards where it wants to be. So simply repeating 'sit down' as a correction for breaking or ignoring 'sit down' will never increase your control. In such cases you need to use something that makes the dog's *final situation worse* than if it had obeyed the first command, such as making it turn around and 'out of it' before making it 'sit down' again, so that the dog views lying down *the first time* as the lesser of two evils.

Intermittent positive reinforcement

If you decided to use positive reinforcement methods to teach a dog something, for example 'sit down', you should start to vary the times when you reward it. That is, you begin by rewarding it with a titbit every time it sits on command, but once it understands the basics you then begin only rewarding it every second or third time it obeys. You extend these intervals in a random manner, rewarding the better responses (and thus correcting the poorer ones). This trains it to obey 'sit' even when a reward hasn't been forthcoming the time or two previously, in the knowledge that sooner or later the reward will come.

This is one force at work in creating gambling addictions in people. They can keep playing for hours with no result or only small wins, hoping for the big one. But they must get some wins occasionally, otherwise they will lose interest. (This principle of 'intermittent reinforcement' has significant implications, and is discussed in detail in the next chapter under the heading of 'Putting two-and-two together'.)

However, training a dog in this titbit manner is very slow and indirect, really only a waste of time. And it will never give you control. You could teach the basic meaning of 'sit down' in this manner if you wanted to, and then gain control by other means, but there are much faster and more effective methods.

Negative reinforcement

Most people assume that negative reinforcement and punishment are the same thing, but this is not the case. Punishment is something undesirable occurring as a result of the dog doing something. So, if the dog stuck his nose in the campfire, it would punish him, and he wouldn't do it again. Or if he tried to take a bone off another dog, the other dog would punish him.

Negative reinforcement involves the *removal of something undesirable* to *reward* the dog when it responds. So you could say that the campfire incident involves negative reinforcement, because when the dog takes his nose out of the fire *the heat stops*, and thus the dog has been *rewarded* for his action.

Punishment is when the fire punishes the dog for sticking his nose in it; negative reinforcement is when he is rewarded because the fire stops hurting him when he takes his nose out.

Negative reinforcement contains both correction and reward. A good example of negative reinforcement is when stopping a horse with the reins. You increase the pressure on the reins, and when the horse responds you ease the pressure off the reins as a reward. You have removed the negative. This is negative reinforcement. You first create a negative, and then remove it as a reward.

Negative reinforcement is often used in dog training. For example, consider getting control of the directional commands. You tell the dog to 'go round' (clockwise around the sheep) and it doesn't respond, but keeps going the way it was going. You should continue repeating and growling 'round' as a correction for not obeying, until you finally make it obey. When it does, you stop growling. So you have rewarded it for obeying by removing the growling.

Summary of positive and negative reinforcement

Basically positive and negative reinforcement are just fancy names for what people have always known – reward, with elements of correction. They both follow similar principles.

The simple fact is that you use reward and correction (and occasionally punishment) to train animals. (Some people like to say that you 'encourage or discourage', but whatever words are used the practicalities remain the same.)

End of lesson rewards

Most rewards, such as those discussed above, are given more or less instantly for specific actions. But there are also broader rewards for less specific and more general improvement.

These rewards are mostly applied by ending a lesson when the dog has improved in the particular aspect that we are working on.

If you keep up a lesson in one particular action too long, the dog begins to think that what it is doing isn't right, because it is being made to do the same thing over and over again. So, once it shows a good level of improvement, you should stop and go on to something different. Much more benefit will be had from putting the dog away for the day, and then when you work it the next day it will have assimilated the lesson, and will probably do it better than the day before.

A good example is when teaching a horse to move its hindquarters sideways in response to pressure from the rider's leg. If you get the horse to understand basically in the first lesson, so that it moves its hindquarters sideways in some way in response, you should accept that as enough and go on to something else, or put the horse away.

If you simply continue with the same lesson, the horse will feel as though it is not gaining anything from moving sideways, because you keep on making it do it again. It is likely to start getting fidgety or maybe kick at your heel, to try to make you stop annoying it.

But if you stop when you see some improvement, and be happy with that, you reward the horse for its cooperation and it will be much better the next day, because it knows that it is on the right track. Next time you will be able to extend the lesson for longer, and ask for more.

So there are small instant rewards, such as easing the pressure with your leg when the horse yields, but then there are also broader rewards by ending the lesson when it shows improvement, before it gets sick of it. If it tires of the lesson and begins resisting, then it is too late, because to end the lesson at this stage would reward it for playing up. This would set your training back.

The reason I have used a horse in this example is because it more clearly demonstrates the various factors involved. But the same applies with a dog, particularly when teaching the directional commands, and when teaching driving away. Repetition is important, but it is better to give the dog a break once it shows some improvement, and repeat it the next day, rather than to continue on and on with the same exercise. You don't have to end the session completely, just change to another exercise.

This is particularly important in the first lesson of anything, and I suggest that you always consider the first lesson to be a sort of 'priming' lesson, and don't try to achieve too much. You may not think that you have made much impression, but it is amazing how you bring the dog out for the second lesson and it generally performs better than it did the day before.

Correction and punishment

Correction is something that many animal trainers don't understand at all (particularly those teaching animals tricks). They don't understand that training with positive reinforcement contains correction (consider the gambler's reaction when they fail to win), nor do they understand the usefulness and indispensable nature of correction.

In fact they often claim that correction or punishment isn't effective, and then use examples of people who know nothing about animal handling to illustrate their point, for example, punishing a dog for eating the garbage hours after it has eaten it! 'See, punishment doesn't work.' (However, some handlers do in fact do this sort of thing, such as by taking a dog off stock and putting it back on the chain as punishment for not performing well, like sending a child to its room. With dogs such measures are a complete waste of time.)

Other claims are sometimes leveled at correction or punishment, such as that they cause 'learned helplessness' or aggression, and so on. Those that make such claims are generally either ignorant of the true facts, or are being very selective in their use of the truth.

Punishment *per se* does not cause these problems, however, incorrect, badly timed or excessive punishment can. It can be argued just as strongly that incorrect use of reward will cause problems. For example, if you reward a dog for timid or aggressive behaviour you will make that behaviour worse, which is often what well-meaning people do. Thus I could claim, in all truth, that 'reward can cause aggression or fear', but this would be a selective and dishonest use of the truth. Reward or punishment in themselves do not cause problems, only their misuse does. When used properly and in context they are both valuable and indeed essential tools.

Correction and punishment follow exactly the same rules as for using reward, *including the use of irregular or intermittent application* (which is a great secret), as discussed in the next chapter under 'Putting two-and-two together'. If these rules are followed, correction and punishment are highly effective. If they aren't, they fail, exactly as a reward does.

The difference between correction and punishment is simply one of degree. Correction is a mild form of rebuke, whereas punishment is something more severe. For example, if a dog failed to sit down when told to do so, a repeated growled command would constitute correction. A shake by the scruff of the neck would constitute punishment.

Where correction ends and punishment begins depends on each dog's temperament. A very hard or excitable dog will require the correction to be more substantial, and punishment likewise. What is punishment to one dog will be only mild correction to another. However, the better your handling and training ability are, the less punishment you will need to use. It should be the exception, not the rule.

Unfortunately, the use of electric collars seems to have become popular in recent years, with their ready availability. They are used by poor handlers *as a substitute for sound training*, and I would never rate any handler as a top handler who resorts to their use. As soon as a handler reaches for an electric collar, in my estimation they have failed. Personally, I have never used one and never will. A good handler has no need for one, and a poor handler doesn't know enough to use one properly.

13
Training tips

So, what is the end result of all this? Basically that we want to create situations in which the dog learns that it gains from obedience, and loses out from disobedience. However, it is a very complex situation with a huge array of variables.

It is all about balance – the stronger a dog's motivation to do something we don't want then the stronger the motivation we must create not to do it. Or we must simply get the dog into the right habit by repetition of what we require, in a controllable situation.

This is where breeding also plays a role. A very excitable, aggressive, rough dog, if it is also very hard in nature, will require considerable motivation to mend its ways. A well-bred dog, calm, steady, and with great natural ability, will pose fewer problems. A soft dog lacking in heart will pose a whole different set of problems, and isn't worth bothering with.

When training, ask yourself, 'Does the dog benefit more from doing what I want, or from what it wants?'

Dogs want to please?

A statement you often hear in relation to dog training is that 'dogs want to please'. First, let me say that this statement is wrong; dogs do not naturally want to please. They want to do what they want to do, what their instincts tell them to do; they want to please themselves.

So to say that if only the dog understood what you wanted, then it would comply, is wrong. A dog may well understand what you want, but doesn't do it simply because it has better things to do.

However, in many situations it is true that the dog simply doesn't understand. Many times when a handler blames a dog for being 'pig headed' or 'stupid' and so on, it is simply that they haven't been clear enough when training it, and it doesn't understand what they want. *In fact this is most often the case.* Usually what the handler needs is a mirror, and he will see the problem.

The dog does what it does (or what it doesn't do) in response to command, because that is exactly the way that it has been trained and handled (or mishandled), or allowed to do. It is nearly always the handler who is at fault.

When a handler can handle soft dogs, but not hard dogs, and particularly at a distance, it is a sure sign that they don't really train their dogs but simply intimidate them, and this is less effective the harder the dog is, or the further away it is.

When training, I realise that when the dog doesn't understand what I am trying to teach it that this is always my own fault for not making it clear enough. Or that when it does the wrong thing, or doesn't do what it is told, that this is always a case of the dog's response being exactly what I have taught it.

I realise that these problems are all of my own making, or due to my own shortcomings as a handler. The hard dogs show your shortcomings up more than the soft ones, and it is from handling those that I have learnt the most.

So I then consider what I have been doing wrong and attempt to discover the reason. I don't blame the dog. So always consider what *you* have done (or are doing) wrong, if you ever want to develop into a good handler.

Most poor handlers tend to punish their dog when things aren't going well, but in most cases this will probably do more harm than good, because the punishment is being used incorrectly.

Timing

'Timing' in the handler is very much like 'short cover' in the dog. A dog with very great short cover adjusts instantly to the sheep's tiniest movements, with tiny movements of its own. Likewise a good handler can read the dog's every thought, and respond quickly and with great tact. The poorer handler takes much longer to pick up on what the dog is doing or thinking, and so responds much later, and generally has to make a bigger response.

If you respond late to correct a dog you can also end up correcting something completely different to what you think you are, and the same when rewarding it. Therefore you must be completely in tune with your dog, in order to ease the pressure at exactly the right time, or to bring more pressure to bear, in order to effectively train it.

One example is if you tell a pup to 'sit down', and then you want to call it off the sheep. You say 'here' whereupon it might look at you, but then look away again at the sheep. When it looks away at the sheep you should growl at it 'youuu', to correct it for thinking about going after the sheep. Then when it looks back at you repeat 'here' in a normal way, to let it know that this response (looking at you instead of the sheep) is a step in the right direction.

By the way you give the command you tell the pup in effect that it is 'getting warmer' or 'getting colder', and so you are explaining what you want, and correcting what you don't. Timing is crucial.

Dog should initiate action

When teaching something new, it is most important for quick and effective results that the dog initiates the response, not the handler.

A simple example is in teaching a dog to sit. Most people push the dog's rump to the ground while saying 'sit' over and over again. The dog eventually learns, but relatively slowly.

Instead, what you should do is apply enough pressure to the dog's rump so that it is uncomfortable and is pushing against your hand to hold itself up. Then hold that position until the *dog sits down by itself*. The dog will then learn rapidly because it sat itself down, instead of just being pushed to the ground.

The same applies when teaching the dog to jump up on to a bale of hay, or onto the sheep's backs. Don't drag it up there, or throw it up, muttering 'get up'. Rather, put it on a lead and hold pressure on the lead towards the bale of hay, and maintain that pressure until the dog *jumps up itself*. Start on a low object and work your way up. The same principle applies in many other areas, such as when teaching a dog to cast.

Some people like to say that you 'give the dog a choice' or 'ask its permission', but that is only playing with words because you only give it one viable option. In fact this is an important principle for good, fast, effective training – you set up a situation where the dog has only one viable response.

The better you do this, and the more obvious you can make this one option, the faster and more easily the dog will learn and understand what you mean.

Many positive reinforcement advocates claim that positive reinforcement training (titbit training) is faster than other methods. In fact this is not due to the positive reinforcement, but because means of causing the dog to initiate the movement are utilised. That is, the trainer sets up a situation where the dog is likely to sit, and then rewards it. So it can be faster than simply pushing a dog into certain positions by force.

But my methods are even faster again, because not only do they cause the dog to initiate the movement, but also they do not give it any other option. Therefore the dog will understand what we want *far* more rapidly, *and we get control*. Positive reinforcement alone will not give control.

Inch by inch

When training, it is important to *gradually* increase the dog's abilities. If you try to get it to do something that it thinks it cannot do, then you can put it off trying even lesser things. If you tried to teach pups to force by working cows and calves, you might put them off working cattle completely.

So start small, and gradually work up. If you want to teach a dog to jump a six-foot fence, start at one foot, and work your way up. To cast a kilometre, start at 10 metres, and gradually work up.

Teach more to achieve perfection at less

However, sometimes you can go against the above rule with good effect, because generally if you have trained a dog to a certain stage it will probably not be perfect at that stage, but it will be at a lesser stage. So if you have taught it to cast a mile it may be reasonable at that, but at half that distance it will be completely confident and competent.

Also, if you have it experienced enough to force cows and calves it may not be 100 per cent confident about that yet, but it will be with less demanding stock.

This is one way of increasing a dog's confidence and ability quickly, provided you are sure that you are not going to ruin it. So to get a dog confident casting 100 metres well, by simply casting it 100 metres, can take some time. But if you go straight out to 500 metres (even if it doesn't do it perfectly), then when you come back to 100 metres it will think, 'There's nothing to it.'

It takes some tact and understanding to know how far you can push this principle, but it can be very useful in quickly training a very well-bred dog.

The same applies to gaining control. If you can control a dog in very trying conditions (such as having it remain lying down while you work a second dog) it will be easy in less demanding conditions.

Another good example is with 'sit down'. Most people only teach the dog to 'sit down' for very short periods of time, and have trouble holding it for longer periods. But if you teach it to stay lying down for half an hour to one command, it will make it much easier to control for shorter periods of time.

It doesn't take much

An important point to realise is that it is remarkable how little work you have to do to make a big difference in a dog. Often by simply giving it only one or two lessons in something you can change it drastically. Many people don't even try because they envisage weeks or months of hard work to achieve something, but this is mostly not the case. One lesson can make a huge and lasting difference. Some things require some repetition, but many things do not, at least not to give the dog the basic idea.

14
Principles of control

I define control as being able to make the dog do exactly what we want it to do, and exactly the opposite to what it is inclined to do, easily and quickly without any resistance even under great temptation.

If you find that you have problems with lack of control in any area, the approach to take is this: work on making the dog do the opposite of what it wants.

For example, if you lack real control of the directional commands, work in a controllable situation and make the dog go in the opposite direction to what it wants to go as much as possible, until you break the resistance.

Ease of control

The better the dog's natural ability, the easier it will be to gain high levels of control of it. And even if it isn't a top dog, certain traits will make it easier to control. First, the good dog is naturally doing the right thing at the right time and therefore less commanding is needed. Less commanding means that there are fewer opportunities for your control to diminish, and less reason for the dog to begin resenting being ordered about. The less commanding you do, the more control you will retain. James Moore was quoted as saying that:

> constant commanding of your dog has the tendency to destroy the control you seek to establish. I would therefore strongly stress the following: – Command your dog as little as possible, but at all times insist that your commands are instantly obeyed, as the sheep dog is quick to take advantage of any relaxation of control. (Moore, in Pearse, no date.)

Second, certain traits serve to make a dog more controllable. A dog that naturally drives sheep along and works (drives/balances) them from behind will be easier to work in contrast to the mad keen heading dog (I am talking about practical work, not trials).

A dog that breaks out to head breakaway sheep cleanly, rather than chopping in, will not need to be constantly roused at to 'get out of it'. A dog that works steadily and doesn't bore in to its sheep all of the time can be left alone to do its work, and will be easier to stop and so on.

Also a dog that is good physically will be easier to control, because it can respond with less effort; it is much harder to get a good 'stop' on a dog which is a big, blundering, poor moving sort of dog than on an agile animal.

So natural ability greatly influences how easily control can be obtained, and then maintained, in practical work. Obviously temperament also plays a big part. The calm, sensible, steady dog will always be superior; you want a dog that thinks before it acts. This type of temperament is essential for high-class working dogs, *but they must be keen*, with great heart.

Certain *faults* can also make a dog easy to control, such as being soft natured or lacking in keenness. Such dogs are undesirable, even though they may be easy to control. If a dog is easily influenced by the handler it will also be easily influenced by a cow with a calf at foot, or by a bad-tempered ram, and won't be much good when the pressure is on.

What is wanted are keen, hard, strong dogs that are very calm and with the right natural ability.

Don't allow the dog to break commands

One of the biggest secrets to gaining control is to never allow a dog to 'break' a command. If you allow it to break a command before it is told you will never gain real control, and what control you did have will diminish.

If you tell it to 'sit down', it should remain there *until it is released*.

Often good control of a dog can be achieved in a training situation, only to have that control disappear on the trial ground or out in the paddock. You may ensure that the dog doesn't break the command at home (but not many handlers do), but when working touchy sheep at a trial you allow it to break 'sit down' before it is told to go, otherwise the sheep may have bolted and points been lost. After a few trials the dog learns that it can get away with this when away from home, and control suffers. The same can happen out in the paddock.

So how do we prevent this from happening? First, we have a *comprehensive* commanding system so that we do not get ourselves into a situation like that above. We have a 'sit down' command that we use when we are sure that this is what we require, but if working touchy sheep we use a different command, 'stop', which tells the dog to stop on its feet but to move when it thinks it is necessary. In this way we have enough commands so that we can avoid this trouble occurring.

If you want good control, *the dog must never be allowed to break the command*. I cannot stress this enough, because it is a mistake the majority of handlers make.

Don't allow the dog to benefit from disobedience

This brings us to the next point. Once you tell a dog to do something, you shouldn't give up until it does as it is told. You often see a handler training a dog and telling it repeatedly to 'sit', while the dog ignores them and continues on its merry way. The handler then gives up and perhaps tells the dog to 'go back'.

What the handler is in fact teaching the dog is *persistence* – persistence in disobedience. The handler teaches it that if it disobeys for long enough it will eventually get its own way. So once you tell a dog to do something, make sure that it does it; never give up.

Another similar tip when training a dog is to make sure that it doesn't gain from substandard work. One good example is when teaching it to cast. The dog cuts in on the cast and is allowed to continue and so is rewarded. Instead you should call it back and make it cast properly, otherwise it has been rewarded for substandard work.

Always remember that working stock has its own inherent reward (if the dog is keen), so that the dog that disobeys is being rewarded by continuing to work the stock.

There are various corrections that you can use in training, and also in practical work. If you tell a dog to 'steady' and it doesn't, make it stop as a correction. It is therefore worse off than if it had obeyed 'steady', because if it had slowed down it would still be working the stock. If it is drawing sheep to you and boring into them and doesn't stop when told, make it turn around and 'out of it' away from the sheep for 50 metres if necessary. Then repeat the whole process (remember repetition and habit) until the dog obeys to a high standard.

Warnings and second chances

Many people are in the habit of giving dogs warnings. The prime example is with 'sit down' (or the equivalent). They tell the dog to 'sit down' and then keep repeating 'sit down' to hold it there, as a warning not to get up and move.

The main reason that this problem has arisen in the first place is that the handler has allowed the dog to *break the command*. So the dog decides when it is going to get up from 'sit down', and hence the handler must keep repeating the command to prevent this. This is the main cause of dogs 'beating the bell' at trials.

Also what they are effectively doing is correcting the dog for lying down. They tell it to 'sit down', which it does, and then keep yelling at it to 'sit down' to keep it there. It hasn't gained much by obeying. It is like calling a dog 'here' and then belting it when it arrives for not coming quickly enough. In both cases control will diminish, more obviously in the latter case, but also in the former.

Another good example is when stopping a horse with the reins. When the horse stops you must slack off on the reins, otherwise the horse has not gained by stopping. If you don't slack off it will try something else to remove the pressure in its mouth, perhaps throwing its head around, or bolting in extreme cases.

It is a similar situation when you tell a dog to 'sit down', and then keep telling it to 'sit down' after it has obeyed. What you should do is give one command and make sure it is obeyed, and then keep quiet and back off completely. Correct the dog (perhaps with a louder command) *if* it goes to move, but never *before*. In fact it is more effective to let the dog just get up before correcting it, because getting up and then being made to lie down again is more effort.

So the rule is: don't give warnings or anticipate disobedience, but simply correct it.

Putting two-and-two together

It is often said that a dog must be corrected for a misdemeanour almost instantly after the misdemeanour has occurred, otherwise it can't associate the two. However, this is not the whole story.

Yes, when a dog is young and is learning something new, correction must follow immediately. But as the dog develops a greater and greater understanding of the action in question this ceases to be so important. It *can* put two-and-two together in *certain situations*.

For example, when first gaining control of 'sit down' you do need to correct the dog for lack of response immediately it fails to obey, perhaps by a stronger 'SIT DOWN'. But as it proceeds in its training, if it takes you a bit longer to get to the dog to correct it, it can learn to associate the two; don't worry about that. As corrections gradually become more delayed during the natural progression of training, so the dog will learn to associate delayed corrections with the previous misdemeanour.

This brings me to my point. If you only ever correct a dog instantly after a fault, and then you get into a less controllable situation such as out mustering with a young dog (or at a trial), and it fails to obey, then when you fail to correct it immediately it effectively says to itself, 'Aha! He is in no position to correct me, I can do what I like!' Which is exactly what he does.

What you should do in order to gain real control is, in a controllable situation, at times *allow the dog to think that it has got away with disobedience, and then correct it*. However, don't allow too long a delay. And remember that this is not something that you do with a young pup, or with a new exercise, or you will create confusion.

This then effectively creates an association in the dog's mind, so that it associates the feeling of 'I got away with that, now I can do what I like', with 'I got away with that . . . hang on a minute, remember what happened last time I thought I got away with that, better do what he said.'

Most handlers will be familiar with the situation where if the dog (probably at a distance) obeys the first time you tell it, then that is good. But if it ignores you the first time repeated commands will probably be of no avail. If you stop commanding for a minute, and then try again, you might then be lucky enough to get a response.

The reasons and cure for this are as outlined above. This is a very important principle, and just one of the things not understood by those with faith in modern so-called 'positive' animal training techniques. These same people often talk about positive reinforcement (reward), and what they call 'intermittent reinforcement'. This means they might only reward the dog every second time, or every third time it obeys a command, but somewhat randomly. In this way they strengthen the dog's response.

Otherwise they would have the same problem as I outlined above, only in reverse. If they had previously rewarded the dog every time it did something, and then one day they had no titbits with them to reward it, the dog decides 'Fine, if that's the way you want it,

I've got better things to do', and loses interest in the 'game' of obedience. When the reward doesn't come immediately it assumes that it is never coming at all.

But if you reward it *irregularly* it continues to do what it has been told even in the absence of a reward (assuming that it hasn't got something more rewarding to do) because it has learnt and hopes that sooner or later the reward will come, just like people addicted to gambling.

Intermittent or irregular correction

What most animal behaviourists fail to realise is that the same principle applies to correction, both in the way described above, which I call delayed correction, but also in intermittent or irregular correction.

So it isn't necessary to correct a dog every time for disobedience, nor even desirable, in a controllable situation. The caveat 'in a controllable situation' is added because in a practical situation you should always try to ensure that the dog does what it is told, probably just with a repeated command or a shouted reprimand.

The main place you utilise these techniques is in a controllable situation. Say, for example, that you are having trouble with the dog not doing what it is told when casting out – not stopping, or not breaking out, etc.

One way would be to put it on a long light rope. Cast it out and hold the end of the rope. As it is casting, sing out to it to 'sit down'. It ignores you and continues on its merry way.

You can then use one of the following techniques. Either delay the correction, so that you allow the dog to go another 10 metres and then use the cord to stop it. Don't allow it to go another 50 metres or it will have forgotten. You don't have to jerk the cord; rather it is more a case of getting the dog into the right habit than of punishing it.

You will teach the dog to associate delayed corrections with earlier commands. So now when the dog gets a command and takes a few more steps, it associates this feeling of having got away with disobedience with 'Perhaps I haven't got away with it', and so the *disobedience itself becomes a cue to obey.*

In this way later on, if the dog for some reason doesn't respond immediately, it may go a bit further but it will then think (in effect), 'I suppose I had better do what he said', and it does. And dogs are like humans, sooner or later they will always test the limits for one reason or another (in fact this is an important part of how animals, and people, learn). But trained in this manner they will usually respond.

The other way is to use 'intermittent correction'. So, with the dog casting out on the long line, give it the first command to stop, which we will assume that it ignores. Then give it a second command, and correct this one. This way, when out working and the dog ignores the first command, subsequent commands are more likely to be respected.

So when training, even if you can't correct a dog every time for something, *it won't know which time you are going to correct it,* and so will improve regardless.

Now all of this ensures that you have 'eventual control'. That is, the dog will do what it is told eventually, and this is very important. But these measures should be the exception, not the rule. The dog should do what it is told quickly the first time.

So in a training situation you should also let the dog know that most times it will not get any warnings or second chances whatsoever. Tell it to 'sit down' once, and if it doesn't then don't tell it again and again and again, but simply correct it, without saying another word. You told it once, that should be enough. It will figure it out.

Now of course these corrections that I talk about vary greatly. A tug on the long rope is one way, or often just a harsh command, all depending on the dog's temperament and level of training, and so on. Generally start with the most softly spoken command possible, and work your way up. Don't always do this, however, or the dog will learn to anticipate. It will effectively think, 'I don't need to do what I am told yet, he's got another four levels of correction before it starts to bother me much.' So vary this also, and sometimes go straight to a higher level of correction. Keep it guessing.

Having said that dogs can put two-and-two together, I want to make it perfectly clear that only a good handler with a clear understanding of the foregoing factors will be able to utilise this fact properly. For most handlers it is better to just work on the premise that correction or punishment or reward must follow the action immediately, even though that isn't strictly true. The problem is that many people give dogs far too much credit for their ability to reason, or to read the handler's mind, and make blunders because of it.

Body language and tone of voice

It is a simple matter to gain 'obedience' (to use the term loosely) at close quarters, particularly for those handlers who use the 'chase-'em-round' method of directional commands, that is who raise their opposite arm to chase the dog around the way they want it to go (this is in contrast to pointing in the direction you require, as many of the top stockmen of the past did).

This 'chase-'em-round' method will result in a sort of half-trained dog very quickly, but it is only the dog responding to body language. Without further work the handler will have no control at a distance as their personal influence diminishes.

Therefore in order to gain control you must eliminate the effects of body language, and teach the dog to respond to the vocal or whistled command itself. You must also eliminate the role of vocal power, that is shouting, in order to achieve obedience, because its influence also diminishes at a distance.

The use of whistles helps to eliminate the role of voice tone. One of the reasons they are generally more effective at a distance is because the sound is similar when given either close, or at a distance. In contrast a shouted command sounds very different to a softly spoken one (in the dog's mind they are really two different commands due to this factor, and the dog needs to be educated to a shouted command as it was to a softly spoken one).

Exercises for achieving this elimination of body language are included in the relevant sections on training each command.

Use a controllable situation

In order to achieve all of these aims it is important to have a controllable situation when training. How controllable it needs to be will depend on the dog's natural ability and temperament, its level of training, and your ability.

One of the cardinal rules is not to give a command unless you can ensure obedience.

So the trick when training is to have a situation wherein you can ensure obedience. Then you gradually increase the level of distractions or temptations. Set up a situation where you can ensure the right response, and then utilise repetition until the habit is firmly fixed. Then gradually increase the level of difficulty. This higher level of distraction may include faster, wilder sheep, or perhaps having the dog remain lying down while working a second dog.

(This is why I suggest not taking a dog out to practical work until it is fully trained, otherwise it will get into bad habits. And, when you do start working it, I suggest working it on its own initially before introducing it to working with other dogs.)

There are a number of ways to achieve this controllable situation. You can train the dog in a yard small enough to enable you to always ensure obedience. Or you can utilise a long, light cord attached to its collar. In practice you will probably utilise one or both at various stages.

With a pup, quiet sheep make a situation more controllable. You can concentrate on the pup and ensure its response, without worrying about the sheep bolting to the other end of the paddock. With quiet sheep you can therefore train in a larger area than you can with wild sheep.

At a later stage, real control of 'sit down' creates a 'controllable situation' anywhere, by becoming a sort of 'invisible rope', and then training is starting to get somewhere. But if you never gain real control of 'sit down' (and most handlers don't) you never reach this basic stage.

Control of 'sit down' is the foundation of all control, because it gives you a controllable situation anywhere.

Summary of gaining control of 'sit down'

There are four main pointers in gaining this absolute control of 'sit down'.

1 Never let the dog break the command.
2 Correct disobedience, don't anticipate it; that is, don't give warnings.
3 Train the dog to stay lying down for extended periods to one command.
4 Utilise corrections other than repeated 'sit down' commands to increase control of 'sit down'.

If you concentrate on these four points you will get control. But if you neglect any one of them you are unlikely to do so.

The use of control

Control is vital in the high quality, practical working dog. But the question is, how should that control be best utilised? The simple answer is *minimally*. That is, as little as possible in order to get the job done to a high standard.

From the emphasis on gaining control in these chapters you may think that the dog is always under control, and never allowed to do anything on its own. But this is not the case. In fact I always strive to work a dog with the least commanding possible, and with a well-bred dog that means very little commanding at all. But when we do give a command the dog should be crystal clear in its mind of what we want, *and of what we are trying to achieve with the stock*, and should obey without hesitation.

When commanding *is* necessary, always use the least 'intrusive' command possible. Therefore use 'steady' before 'stop' before 'sit down' (depending on what you want, of course). Often with a fully trained and experienced dog a single short whistle (or its name) is enough to tell the dog that what it is doing is not what you want, so it should do something else. The dog that has enough training and experience will know what that something else is.

So although you have the dog trained to a comprehensive commanding system (as explained in Chapters 16 and 17), that training then allows you to come back to a very simple way of working it. But as soon as things get interesting you have your whole repertoire of tricks to fall back on.

Allow the well-bred dog to do as much as possible on its own initiative. The more you can leave to the dog the better (while still achieving a high level of stock work), and the more control you will retain for when you really need it. But if the dog is not handling the stock in a sound manner, use your control to make it do so. The dog is there to assist in the sound management of the stock, not the other way around. *No matter how much natural ability a dog possesses (and I am as great an advocate of natural ability as anyone), you must still have the dog trained and under good control.*

In the Kelpie breed in modern times almost a tradition of lack of control and lack of training has evolved, through such comments as 'Kelpies are do-it-yourself kits, not push-button dogs'. Such statements then tend to become cop-outs, or self-fulfilling prophecies, when handlers run into difficulties gaining control of, or training, their dogs. They blame their lack of control on the dogs, and use the excuse that 'Kelpies aren't push-button dogs', and so give up easily. Yet the best Kelpie handlers of the past had high levels of control of their dogs, as do the better ones today.

Just because a dog does what it is told does not make it any less of a natural worker, or any less of a thinking worker working on its own initiative, if it is allowed to do so. In fact high levels of control and training, and more comprehensive commanding systems, allow a dog to do more work on its own initiative, and to a higher standard, than otherwise (this is discussed in Chapter 16).

A final point, also to be discussed in Chapter 16, is your commanding system. If you have too few commands, so that the dog is unsure of what you want, then the situation is less than ideal. Imagine yourself working for a boss who doesn't speak English. He only has one word that you both understand for 'tool', and yells at you to get the 'tool', but you haven't any idea which tool out of hundreds he is talking about. You wouldn't be feeling too cooperative as he keeps yelling at you every time you bring back the wrong tool.

But this is exactly the situation with handlers who use simple commanding systems with *too few commands*, and then yell at their dog when it doesn't understand what they want.

A high level of communication and control are necessary (because it is no good being able to tell a dog what you want if it ignores you), and this will enable you to teach the dog practical work very quickly. You can then work the dog in most cases with very minimal commanding (if it has the right natural ability). But when tricky situations arise you have developed a system of communication to the point where you can tell the dog exactly what you require.

15

The long rope

There are various methods of training dogs. Some people argue that you shouldn't use a long rope, but do all of the training without it. Some say that you should train completely in an open paddock, and not using a yard. Others do nearly all of their training in a yard.

Some people say that training a dog is a four-year job, others say that it takes two years. Others only take a couple of months, but these are generally only training their dogs for simple trials (to simple, mechanical commanding systems), or aren't really training their dogs at all.

Some people say that a year in the dog's life is worth seven in a human's, so that a three-year-old dog is the equivalent of a 21-year-old person. But this is wrong. A dog is fairly mature at 12 months old, much more than the equivalent of a seven-year-old child. The 12-month-old dog is more like an 18-year-old, and is ready to start serious work at that age.

The better your training and handling techniques are, and the more closely you follow sound training principles, the less time it takes to train a dog.

I originally started training dogs completely in the open paddock, with no recourse to a long rope or a yard. You can train dogs like this, and with some dogs it is very easy and effective. The UK dogs are the best example. Many have exceptional temperaments and great intelligence, and are very forgiving. Australian dogs generally have worse temperaments and pose more problems to train and handle.

The worse the dog's temperament is, and the harder it is to handle, the better your handling techniques must be to get good results. It is not that you need to be a *harder* handler to handle a hard dog, but that you need to be a *better* handler in all regards.

I learnt more about handling dogs when training Moorlands Georgie than any other dog I have owned. She was exceptionally hard to handle, but I persevered because of her great natural ability in many areas (if she had been perfect she wouldn't have been so hard to handle), and learnt. And no matter how naturally gifted a dog is (and she was exceptional) you still need control, and more so the more difficult and varied the conditions are.

The long rope – and the *very* long rope

She taught me, among many other things, the great value of the long rope. My good friend, Sid Cavanagh, makes use of the long rope, and I learnt a lot from him.

Without the long rope you simply train a dog in everything close at hand, and then gradually, step by step, increase the distance until you have control at any distance. But with most dogs it is a slow process. However, with the long rope you can increase the distance that you have control at in leaps and bounds, and training progresses much faster.

Also if you can give the dog only one option when teaching it something, it will quickly realise what you want. Alternatively, if it has a number of options that it can try it will learn much more slowly. So the long rope allows you to teach new things, such as driving, and breaking out on the cast, very rapidly, because it prevents the dog from *ever*, right from the start, doing anything wrong, and limits its options. This greatly accelerates training.

I wouldn't train a dog nowadays without a long rope, knowing what I now know. You can train a dog in a fraction of the time that it takes without it. I would say that with a well-bred pup I can have it fully trained in no more than six months. By fully trained I mean to all of the directional commands, driving away, splitting sheep, casting, recasting, and so on, as explained later. With a poorer pup it will take longer. If you are only training to a simple (mechanical) commanding system, for nothing more than three-sheep or yard trial show work, it takes less time than for high-quality practical work.

My advice to any dog handler is to learn about the magic of the long rope. It is also very useful for the general farmer – two common problems they have with young dogs are trouble calling them off, and trouble keeping them back driving with them – the long rope is the easiest solution to both.

16

Commanding systems and their role in achieving control and initiative

NOTE: The commanding system described in this book is based on the system devised by Rod Cavanagh, author of *Australian sheep dogs – training and handling*. However, there are some slight differences, the main one being that I swap the roles to some extent of the two commands 'come' and 'here'. I believe that all practical dog people are indebted to Rod for developing and outlining this system of commanding.

The roles of a commanding system

As hinted at in the preceding chapters, the commanding system used is vital in easily achieving high levels of control. In fact I would go so far as to say that it is nearly the *single biggest factor* in achieving that aim, for practical work.

But not only is it vital in achieving high levels of control with ease, *but it is also the prime factor in achieving a dog working as much as possible on its own initiative, and with great purpose.* This is something that you won't learn about from trial handlers (either here or in the UK) who are only interested in mechanical dogs doing show work.

So when considering which commanding system to use, you should keep these two factors in mind:

1 control of the dog
2 initiative and purpose in the dog's work.

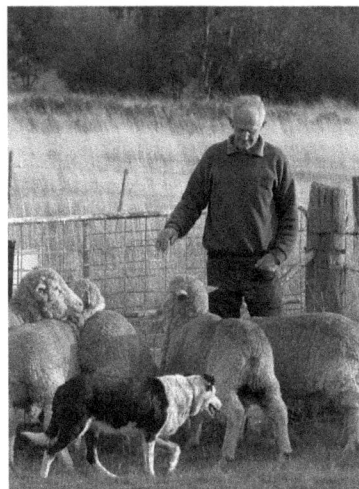

Rod Cavanagh with Cullen's Penny
(Barravore Jim × McIntosh's Meg)

Control of the dog

In order to achieve the first aim quickly and easily, that is, real control, a commanding system must have the following attributes:

- enable us to tell the dog specifically what we want it to do
- to be clear, with no overlap or contradictions between commands
- to help prevent the dog from breaking commands.

I will examine each point in turn. First, any commanding system must enable us to tell the dog specifically what we want it to do, such as stop or move clockwise or anti-clockwise around the sheep, and so on. Remember the boss who only had one word for 'tool' – you need enough commands so the dog knows *exactly* what you mean.

Second, we must not have any overlap or contradiction between commands. So we wouldn't use 'here' both to bring the dog to us, and to send it away as one of the directional commands. This is a simple example, but the principle is important in many areas and many people break it. This is mostly because they use a commanding system that has evolved with no real thought rather than being consciously designed with these points in mind. *Each command should have one, and only one, specific meaning.*

And finally, the commanding system should help us in preventing the dog from breaking commands. How it does this has been touched on earlier when talking about the desirability of having more than one stop command. If you have only one stop command, and you use it sometimes to just pause the dog standing up and at other times to make it lie down and stay there, then you are going to have trouble.

Thus having more commands (which ties into the first point of having enough commands to tell the dog specifically what we want it to do) helps us to prevent the dog from breaking them.

Initiative

In order to develop a dog that also works with the greatest possible initiative and purpose, we need a commanding system which tells the dog not only *specifically* what we want it to do (that is, go right, go left, stop, and so on) but also that lets it know *generally* what we want it to achieve, such as to drive the sheep in that direction, or to bring the sheep to us, or to leave the sheep it is working and look for others. Thus a good commanding system must be both SPECIFIC, and GENERAL.

So, in order to gain *purpose* in the dog's work, we need other commands that tell it what we want it to do with the sheep in a broad sense – thus 'go back' tells it to go around to the far side of sheep (without crossing) and work them back to us; 'go forwards' when driving tells it to go up to the lead and turn the mob without heading it, and so on. So we have commands that tell the dog what we want it to do with the mob.

These extra commands also help to prevent the dog from breaking commands. For example, another area most prone to problems with commands being broken is the directional commands. This is an area that many people have trouble getting control of.

They give a directional command but then allow the dog to decide *how far* to go in that direction. They have no way of telling it how far to go. This means that in some situations they struggle to get it to go far enough, and in others they struggle to prevent it from going too far.

Now there are a number of ways to rectify this problem. You can do it either with a simple commanding system, or with a comprehensive one. The way some handlers get around this problem is that they tell the dog to go in a certain direction and don't allow it to stop until it is told to. Thus they don't allow it to break the command.

It results in high levels of control if done properly, particularly with a certain type of dog. But it results in mechanical work because the dog never knows how far you want it to go, *and thus never knows what you are trying to achieve with the sheep.* It just does what it is told.

The other way to fix this problem is by a more comprehensive commanding system. So to our two basic directional commands ('round' for clockwise and 'over' for anti-clockwise) we add further commands to more specifically tell the dog what we want.

We add 'go back' to tell it to go all the way around the far side of sheep and work (pull) them back to us. Or we tell it 'here' to come from the far side of the sheep around to our side, now to drive sheep. Or, if the dog is driving sheep with us, we tell it 'behind' to bring it around behind us to drive along on the other side of us.

Imagine yourself and a young dog tailing a mob of sheep along a road, using a simple (mechanical) system of only two directional commands. If you want the dog to go in one direction you might tell it to 'get over' (anti-clockwise). But what does your young, over-keen heading dog do? It goes 'over', but it keeps going 'over' and runs all the way around the mob to the lead and blocks them. After much trouble you get it back to your side again.

After a bit of work the dog has just about got out of this problem, and in response to 'get over' it doesn't rush to head the mob, but remains driving them as you desire. Then one day you are again droving a mob along, and the sheep bolt down a wrong road, and you tell your dog to 'get over' to head them. But it doesn't. You have taught it not to do that. So the mob disappears down the road with the dog chasing it away, and you yelling at the dog to 'GET OVER'.

This illustrates how a simple commanding system won't give control, if the dog chooses when to stop going in that direction. With a simple commanding system the way you can gain real control is not to allow the dog to break the command. So you tell it to 'get over', and you must make it keep going 'over' until you tell it to 'stop'. So when droving the mob along, every time you tell the dog to go somewhere you must then 'stop' it where you want, and not let it stop on its own. Then if you want it to head the sheep it will readily do so when you don't stop it.

Then there are other handlers who just keep repeating the command until the dog is where they want it, but again this is less than ideal. The dog still has no initiative. It does no thinking for itself. It *cannot* know what you are trying to achieve.

Thus we must add more commands. So we reserve the two directional commands for going in either direction, but we teach the dog that in response to these directional commands it must move in the required direction, but only so far so that it *remains driving away from us if it is driving away, or remains pulling to us if it is pulling.*

In this way the dog knows its role, and what we are trying to do with the stock.

Then we add 'go back' to tell it to head the mob when that is required. We also add 'here' to tell it to return from a pulling role to a driving role. We add 'behind' to tell it to come towards us and continue around behind us, to our other side, when driving.

Thus we can not only tell the dog which direction to go in, but also how far (or how far not) to go in that direction. So the dog now doesn't break the command (not to any practical extent). And we have achieved the other great point in gaining initiative: we tell the dog what we are trying to achieve with the stock.

Telling the dog what we are trying to achieve

The benefits of this are that the dog now knows when we are driving stock away, and it knows when we are pulling stock. It knows if we want to turn the mob but not block it, 'go forwards'. It can now develop purpose in its work, and very quickly, and yet we retain high levels of control at the same time.

It is confident in its actions because it knows exactly what we want and what we are doing with the stock, and it doesn't have to guess or try to read our mind.

A good analogy of the two types of commanding systems is as if you are trying to tell a person to walk over to a certain tree. You could tell them to 'pivot right … stop there (when they are facing the right direction), walk forwards … stop there, pivot left … stop there, walk forwards … stop there', and so forth, until they were standing at the tree you wanted. This is how simple commanding systems work. You tell the dog specifically what to do, but it has no idea of the long-term purpose of its actions.

In a comprehensive system with general commands you would in effect simply say, 'walk over and stand near that tree', then leave it to the person to figure out how to do that. This is what happens with the dog in a comprehensive commanding system.

You tell it what you want done in a general sense, but then allow it to figure out how to achieve it.

Many people think that the fewer commands you have then the more the dog is working on its own initiative, but this is the opposite of the true situation. Also consider that there is *purposeful* initiative, and *purposeless* – an untrained dog is working on its own initiative, but as far as we are concerned it is purposeless. Only with training, and communication, does it come to have *purposeful* initiative.

'Get over' or just 'over'

We can further refine our control of the directional commands, by using 'get' as a prefix to the directional commands when we require a large movement, thus 'get over'. For a smaller movement we just use 'over'. For a smaller movement still, we may only say the first part of the command, thus 'ohhh'.

Summary

As outlined above, comprehensive commanding systems are the only viable method of commanding *practical working dogs*, if we want to achieve both high levels of control and high levels of *purposeful* initiative.

Indeed we cannot gain purposeful initiative without high levels of control, and certainly not quickly. A dog may develop reasonable purpose in repetitive tasks such as bringing the dairy cows in, over years of farm work with only an average handler, but it won't be much use in other situations. But, with a good handler and trained to a good commanding system, a young dog develops great purpose and knowledge of all stock work in the mere space of a few months, or even less, and understands what you are trying to achieve even in unusual circumstances.

Even if no other action is taken by a dog handler than to change to a well thought out, comprehensive commanding system, they will see great improvements in their dogs, both in control and in purpose. When this is combined with great natural ability, the sky is the limit.

17

Overview of commands

So, in this comprehensive commanding system we have two basic directional commands, 'round' (clockwise) and 'over' (anti-clockwise). We can add 'get' if we want a bigger movement, or shorten the command if we want a very small movement. The dog remains driving or pulling, whichever it is doing, unless another command is given that tells it that we now want it to change what it is doing with the stock. For this we have various add-ons including 'back', 'behind', 'here', and 'come'.

'Back', either on its own 'go back', or in conjunction with a directional command 'round back', tells the dog to go in that direction all the way around to the far side of the mob, and to work them back to the handler. *The directional command tells the dog which way to go, the add-on tells it how far, and what to do with the stock.*

'Behind' is used when driving stock to bring the dog around behind the handler out onto their other side. It can be used on its own, or with a directional command, 'round behind'. The directional commands tells the dog which way to go, the 'behind' tells it *how far*, but that we are still driving the sheep.

'Here' calls the dog from a pulling role on the far side of the mob, back around to our side to a driving role. It can also be combined with a directional command, 'round here'. The directional command tells it which direction to go, 'here' tells it *how far*, and what we are now trying to do with the stock, that we are now driving them.

'Come' tells the dog to cross *between* us and the sheep, in contrast to 'behind' which tells it to change sides *behind* us. 'Come' is also used to bring the dog from a pulling role to ring the sheep (cross) and drive them away. It is used in conjunction with a directional command, such as 'come round'.

Arm signals

The way I use the arm signals is different to the popular modern method. I raise my arm and point in the direction that I want the dog to go. This is in contrast to most people who raise their *opposite* arm to chase the dog in the direction they want it to go, or to block it from going in the direction they don't.

Many of the best stockmen of the past used the former method of pointing in the direction that they wanted the dog to go, and I far prefer it. Jack Hiscock used that method, as does Rod Cavanagh. I have seen photos and film of Frank Scanlon working Kelpies, and it is evident that he also used that method.

Some people use a mixture of the two methods, which is even worse.

You must be consistent, one way or the other.

You can train a dog to either method of arm signals, but I prefer this method. I find that the dog's attitude to the handler is poorer in the 'chase-'em-round' method, because it is taught to always run from pressure applied by the handler. Also it is more difficult to get control of 'sit down' or 'stop there', because any time the handler goes towards the dog it runs away, because it has been taught to always run from pressure.

Also the 'chase-'em-round' method tends to cause the dog to break wider off its sheep, and the bigger natural casting dog will often break too wide, particularly when coming around behind you. Also it is difficult to give the dog clues as to the direction of stock when out mustering with that method (this is discussed in more detail when discussing teaching the dog to cast and muster). It can also tend to make an excitable dog worse.

That method is popular because it is quicker initially, but for the high quality mustering dog I believe that it is better to point in the direction that you want the dog to go. However, it is difficult to change a dog over to this system if it has previously been trained with the 'chase-em-round' method. Therefore, if you do want to try this method, I suggest starting with a fresh pup and you won't have much trouble.

To use these arm signals, you face the dog and point out to the side in the direction that you want it to go (with the right arm for 'round' (clockwise), and the left for 'over'). If you have your back to the dog you use the opposite arm.

Commanding system summary

Inhibiting commands

'Sit down'	Dog lies down and remains there until told to move.
'Stop there'	Dog stops standing up, and remains there until told to move.
'Stop'	Dog stops standing up, but is permitted to move and cover.
'Steady'	Dog slows down.
'Stay'	Remain in one area, such as back of ute, etc.

Directional commands

'Round'	Go clockwise around stock. Remain either driving or pulling.
'Over'	Go anti-clockwise around stock. Remain either driving or pulling.
'Get round'	More definite/larger/more urgent movement required, but still remains driving or pulling.
'Get over'	More definite/larger/more urgent movement required, but still remains driving or pulling.
'To me'	Come directly towards handler when driving.

Positioning commands (in either direction)

'Back'	Sends the dog around to the far side of sheep without crossing, to work them back to the handler. Used on its own, or with a directional command, 'Round back'.
'Behind'	Pulls the dog around behind the handler to their other side. Dog to remain driving and not continue around to head. Used on its own, or with a directional command, 'Round behind'.
'Here'	Used to bring the dog from a pulling position to a driving one, either on its own or with a directional command, 'Over here'.
'Come'	Used with a directional command to pull the dog around between handler and sheep, to cross, and drive the sheep away.
'Go forwards'	To send the dog to the shoulder of the mob to turn it or string it along, or when working a drenching race or backing.

Casting/mustering commands

'Go back'	To cast around stock and work them back to handler.
'Go out'	Bigger, wider, searching/mustering cast.
'Get out'	Break out and widen the cast.
'Come in'	To cut in and narrow the cast.
'Get away'	To leave sheep that the dog is working and recast for others further away.
'Here away'	To leave sheep that the dog is driving and come back towards or behind the handler to retrieve or recast around sheep.

Forcing commands

'Walk in'	To walk in steadily and directly towards sheep and force them.
'Push in'	More vigorous, free-moving force.
'Speak'	Bark.

Wide working commands

'Keep off'	Walk backwards.
'Out of it'	Turn and go away from stock.

Miscellaneous commands

'Here'	Come right up to handler.
'Heel'	Follow behind handler until told otherwise.
'Hop up'	Jump up on sheep's backs, or ute, etc.
'Hop down'	Jump down onto the ground.
'Quiet'	Stop barking.
'Split 'em'	Divide mob.
'This one'	Cut one sheep off.
'Ahhh', etc.	General reprimands.
'Good dog'	General praise.

Part 4
TRAINING

18

Rearing

The early months of the pup's life are very important, because this is when much of its character development and growth occurs, and also when all sorts of bad habits can form, if allowed. Care at this stage will be well rewarded – it is far easier to train an unspoilt youngster than it is to re-educate a spoilt one.

The pup must have plenty of freedom, to encourage its confidence and intelligence to develop, but this must be somewhat limited so that bad habits are not allowed to form. It should not be allowed to wander off and find stock to work on its own, as this will quickly become a bad habit. Also be careful not to give commands that may not be obeyed, because that would only teach the pup to ignore them.

Tie up

Anytime from about three months onwards you can teach the pup to tie up. Put a leather collar on it, and leave it on a day or two. Then attach the pup to a dog chain and let it drag that around for a bit. Next put it on a lead of about 5 metres long and take it for a walk. This is kinder than simply tying it up without any preliminary work, depending on the pup.

Then tie the pup up somewhere where it can't get tangled up.

The pup's reaction to being tied up the first time is a good indicator of some

A four-month-old Kelpie bitch pup tied up for the first time

aspects of its temperament. The better pup won't panic – it may struggle a bit initially but will *very* quickly accept the situation, even without any preliminary work.

Evaluate

The next step is to decide whether or not the pup has enough potential to be worth persevering with. There is no sense in putting a lot of time and effort into a pup that lacks natural ability. Breeding and natural ability are vital, and without these two ingredients success at high-quality practical work is impossible.

There are a number of stages of this evaluation. The first stage is the pup's temperament; the second stage is its ability to hold a few sheep up against a fence; the third stage is its ability to work sheep in the open; and the fourth stage is its ability to cast naturally and to naturally pull the sheep back to the handler.

Evaluating temperament

First, you can roughly gauge the pup's temperament at a young age, often as early as four or five weeks. Ideally it should be very calm and relaxed (and yet free moving – not doughy) and very bold. CALM and BOLD. *Timid, excitable or soft pups are undesirable.*

A good test is to pick a pup up. It should remain relaxed and unconcerned. Pups that wriggle and squirm are generally excitable and busy, while those that tense themselves or hold on tightly are generally frightened.

A good way to pick a bold pup is to make some loud noise or disturbance, such as rattling pebbles in a tin can, or clapping your hands, or hitting a tin wall, or cracking the stockwhip, or waving a bag around.

I also like short-haired pups because long hair is nothing but a nuisance in Australian conditions with grass seeds. No one serious about breeding working dogs in this country should be deliberately breeding long-haired dogs – they have no advantage over short-haired dogs, and a number of disadvantages. (However, I will put up with the inconvenience of a long-haired dog – if it is so good that it would be foolish not to – but only in very rare cases.)

Also the pup must have good pigmentation around its eyes and nose, otherwise it will suffer from sunburn. Any fairy-tales about the colour of a dog's eyes, and so on, can safely be ignored.

Evaluating the pup on sheep initially

Next we must evaluate the pup on sheep, from when it is about four or five months old, and there are a number of reasons why people commonly make mistakes at this stage. One is that they try the pup on quiet sheep in a small round yard, where it is impossible to see what is in the pup. You must try it in a big yard or a small paddock, with free moving sheep.

Also most people simply don't know what to look for (this subject was examined more fully in Part 1, 'Natural ability and selection'), and most people's standards are far too low. The other main reason for mistakes is that handlers and breeders make excuses for the pup based on its age. *A dog's natural instincts will not change over time.* The

A Campaspe bitch pup (Barravore Jim × Boho Misty) has a first look at sheep

instincts that a pup demonstrates (once it is keen enough to work) are the instincts that will stay with it for life.

The only allowance you should make for a pup is that it will get faster with age (and even that isn't as important as many people believe). Excuses such as 'He's going through that puppy stage' are without foundation – if the pup is good enough it will always be good enough, right from the start.

The principle of evaluating pups is simple: look for the old dog in the young pup. You shouldn't have to teach it *how to handle stock*. You should only have to establish a means of communicating, and gain control.

The pup shouldn't do anything wrong when it is given complete freedom to do its own thing with sheep. And I don't mean with a few dead quiet sheep in a tiny round yard. I mean with free moving sheep in a good-sized yard or half-acre paddock. The size of the enclosure is only limited due to the pup's lack of speed.

However, a good pup doesn't need much speed, because it doesn't cause sheep to panic and split or bolt, and because it breaks off them when they run, and gets fully to the head to turn breakaways back. It is surprising how a good pup can handle sheep faster than itself in a sizeable area, whereas a poor pup will fail even with quiet sheep in a small area.

Don't attempt to influence the pup yourself, by standing between it and the sheep and using plastic rakes or such like to keep it away from them. Just sit on the fence and observe. If you are worried about what is going to happen then you might as well not bother, and start looking for a better pup right away.

So what does 'handling' sheep entail? First, what it does *not* entail. It does not entail splitting the sheep up, or biting or grabbing hold and hanging on, or showing fear of the sheep, or turn tailing. Any of these things are enough reason to be sure that the pup is not of high quality.

What is wanted is a pup that keeps out off its sheep and holds them all together calmly without any training whatsoever. This means that when the sheep move the pup covers their movement, while maintaining its own distance from them.

If a sheep breaks away and bolts, the good pup will get wider as it runs to head it off until it gets in front of it. This is one reason why a good pup can handle even faster sheep. The good pup breaks out off the breakaway *until it turns it back*. A poor pup will chase along behind or

A Campaspe dog pup (Barravore Jim × Boho Misty) showing interest

shoulder the breakaway, and only make the situation worse, or will even cut between it and the other sheep. *A pup should always maintain its own side of any breakaway sheep.*

Generally the good pup will get the sheep under control and held on a fence, and will then show its force by walking in directly to shift them, covering them while it does so and breaking out to head any that break away. You can also try it on a mob of 100 or so in a corner or against a fence, and see if it keeps out around all of them and holds them together.

The main things I look for are a *calm* pup that *breaks out* to head sheep cleanly, and that gets fully in front of the head when it does so to *turn them back*, and is *even* on both sides; it should have good *short cover*, and show *strength* and a desire to *force* without rushing or diving in, and some signs of *drive*.

You can downgrade most pups generally being bred on this test alone. But this is only the first step.

Evaluating on sheep: 'go back'

If the pup passes this test, you then need to wait until you have broken it in to 'go back' to go around the far side of the sheep to hold them to you.

The good pup will very readily understand to go around behind sheep in this manner, and will almost break itself in, even out in the open. With a poorer pup you will have more trouble, and may have to take it into a smaller yard where you may still have trouble. Even keen holding dogs like Moorlands Georgie, or Cummins Creek Chocker, or Campaspe Bounce, inherently understood to go around behind sheep in a tight corner without simply holding them into it, and to pull them back to me. This shows that they had the idea of moving stock, as well as heading them. Most dogs have no idea. If a pup doesn't have this ability it is very unlikely to ever cast well.

Only then, once the pup is broken in to 'go back', can you see how it reacts when working a full circle around sheep in the open, rather than only in a semi-circle against a fence. Only then do you get a really good idea of its width, square break, balance, drive, distance, and hold together.

The pup should arc out around the tail of the sheep (width) when it is drawing them to you, and not cut straight across it. Also it should balance the sheep to you without too much moving about, rather than rushing from one side to the other looking for the head on each side.

Evaluating the natural cast and draw

Even then we still don't know if the pup has real casting ability or a natural pull (although we should have some idea). So take the pup out, once it understands 'go back', and start sending it longer distances. See if it will quickly come to cast wide and deep a couple of hundred metres or more, and if it will work the sheep back to you with no commanding, without ringing the sheep or any such carry on.

This subject was discussed in Chapter 5 in much more detail.

Summary of evaluation

The simple rule is this: *look for the old dog in the young pup.*
The stages of evaluation of a pup are:

1   Temperament as a young pup.
2   When first let loose on sheep, once it is keen and working.
3   After you have broken it in to 'go back'.
4   Testing its natural cast, lift and draw.

'Here'

'Here' is a good first command to teach. This tells the pup to come all the way to you. You can introduce this command if the pup is coming to you of its own volition: sing out 'here' as it is about to come, and praise it when it arrives. This lays down an early association.

Often, at about four months of age, pups can become cunning in avoiding going into their yard or onto the chain. This is more likely in a pup kept on its own, because otherwise the competition for food between litter-mates provides the incentive for them to return to their yard. It is also more likely in soft or timid pups. If a pup does become cunning to yard, it provides a good opportunity for training.

Any situations in which a dog or pup does not want to obey, or wants to actively disobey, provide the best training situations – provided that you can ensure that it does obey. Indeed, the greater the pup's resistance, the better it will learn the lesson, and the more control you will obtain.

So put the pup on a light cord about five metres long. Move towards the dog yard, and command 'here', followed by a tug on the cord if the pup is unresponsive. No other extraneous 'commands' should be given (some people have a habit of talking all kinds of rubbish). Give the command, and that is all. No 'Come on, big fella, what are you doing, come over this way, HERE'. Just plain simple 'here', once.

If the pup fails to come right up to you, or attempts to move away, call it again and repeat the tug. Note that 'tug' does not mean 'pull' – you must tug, then immediately slack. TUG and SLACK. You are not *pulling* the pup to you; *rather you want it to come to you on its own.*

Very quickly it will realise that the wise course of action is to come to you. When it obeys give it a pat, then walk away and repeat the exercise. In this manner the pup is learning 'here'. It is pointless simply reeling it in on a constantly tight lead while nagging away, 'here, here, here'; all it will learn is to pull against you.

What you must achieve is the pup *coming to you*, not you pulling it to you.

A situation where the pup is resisting coming to its yard is the ideal set-up for teaching obedience. In other situations where the pup is *wanting* to come to you, you can instil some idea of the command's meaning, but you can never gain control. You will only gain control by enforcing a command when the dog is otherwise inclined.

For example, another method is to give the pup a meaty bone, call 'here', and proceed in the manner just described. Later on, the sheep provide the resistance to train against in the same way that the meaty bone does, or in the same way that the pup not wanting to come to its yard does.

Extended 'here' method, and the timid dog

The method I am about to describe is based on the idea of setting up a situation where the dog is averse to obeying 'here', and then showing it that obedience is the best (and only) option.

This method will do great things for timid dogs – vastly improving their disposition. They will gain a lot of confidence in you and also in themselves, contrary to how it might seem at first. By timid dogs overcoming their fear they discover that the situation caused no harm and they thereby master their fear. They will be more self-confident and have more confidence in you.

The first step is to have the pup on a lead and basically do the previous 'tug and slack' exercise, *but moving directly towards the pup and growling 'here' as you do so*, so that it is inclined to run away from you. This is calculated to make it exactly *not* want to come to you.

Repeatedly growl 'here' while continuously moving towards the pup, and give a tug any time it turns away. When it comes to you, immediately drop all intimidation and switch to the kindest possible praise, and really make a fuss. This is an example of 'negative reinforcement' in practice – you create a negative and then remove it when the pup obeys, and use strong praise.

Repeat the exercise once the pup moves away. It will soon learn that the best place is close to you where it receives a great fuss and praise, even when it doesn't want to come.

Then you do the same thing, only while waving a bag or a leafy branch in your free hand at the same time. Stand against the wall of a shed or against a fence, and wave the bag in the air somewhat in the pup's direction, while commanding 'here' in a gruff tone of voice. The reason for standing against the shed is to prevent the pup from simply circling around you. Repeat the previous exercise while constantly waving the bag. Do not completely block the pup's path to you with the bag, or it cannot possibly obey.

This technique for teaching 'here' is supremely effective for improving the timid pup. It learns that if it comes to you, all of the bag waving and so on stops. This is what does such wonders for it – it braves something that it is afraid of (the waving bag), which then stops. Running away doesn't benefit it, braving it out does. Its confidence rises greatly.

I once bought a very timid dog about 18 months old. She wouldn't come within 20 yards of her owner when I bought her – not that she had been mistreated, but she was just naturally very timid. After this lesson she lost all of that fearfulness and was a different dog altogether. (However, I don't recommend keeping dogs with any sort of timidity or softness. They are never much good.)

This is the same principle as one method of breaking a young horse in. With the horse on a lead rope you flick a bag around and on it. The horse will initially jump around and try to get away from the bag, but you just quietly persevere until it stands still, even if only for a second, when you immediately stop waving the bag as a reward for standing still.

Then you repeat this until the horse views any frightening object as a cue to stand still. Then when you put a saddle on it, and finally yourself, you have no trouble. Also the horse develops trust in you and becomes much calmer and less timid, and loses its fear of all these sorts of things.

This is also the same principle at use in the so-called 'horse whisperer' techniques, and methods such as the Jeffrey method of horse breaking, and other much older (and sometimes more effective) methods that mostly follow the same principle, just in different ways. They are nothing new. Most people don't understand the reason why these methods work (even those teaching them) – and all kinds of outlandish explanations are put forth, but they all follow the same basic principles.

The author with a young stock horse filly

'Chase-'em-round' arm signals

One problem with the fashionable 'chase-'em-round-with-a-rake' method of training dogs is that it works completely opposite to these principles. Instead of rewarding the dog when it comes towards the pressure and braves it out, you reward it when it runs away from the pressure.

This will tend to make a timid dog worse, and is not an ideal way to develop a good relationship between the handler and *any* dog.

19

Early sheep work

Up until this point the pup has been given a free run on some sheep every now and then, for the purposes of developing its keenness and evaluating its potential. During these times no commands should have been given, *because to give commands before they have been taught will only teach the pup to ignore them.*

We haven't been trying to influence the pup, but we simply sat on the fence and evaluated its potential, and allowed its keenness to develop. So, having decided that it is worth persevering with, we are now ready to start breaking it in.

I like a pup to be at least six months old before I start, so that it has some maturity and speed.

'Go back'

The first aim is to teach the pup to stay on the far side of our small mob of a dozen or so quiet sheep, and so to bring them along behind us wherever we walk. The best place to do this is in a sheep yard about 10 metres by 10 metres.

I never use a round yard, and can't see any value in one whatsoever. A reasonable-sized square yard is better, because the pup learns to get sheep out of a corner and not to cross, and it can get off the sheep a bit and balance and hold them. At a later stage, a situation with corners and kinks is even better.

So take the pup (on a light cord three or four metres in length) to the yard, and let it go free with the cord dragging. You shouldn't need to keep the pup off the sheep, to prevent it splitting them or grabbing hold, because if it does that it will never be a top dog.

However, because most pups lack much ability and split the sheep up and so on, to keep out off the sheep is about the first thing many handlers teach their pups. To work wide is also one of the early things many trial handlers teach. But with a natural pup for

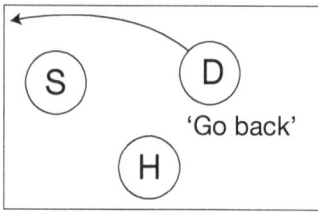

Figure 19.1 'Go back'

practical work I let its force develop first, and don't teach wide work until much later.

If the pup does split the sheep, you will have to find some way of keeping it out, perhaps by getting in its path and waving a bag or a leafy branch at it. At the same time growl at it, until it understands. However, if it is a soft pup this might put it off work altogether, so with a poor pup it can be a fine balancing act in the early stages.

Gradually begin to position yourself to block the pup from holding the sheep against the yard fence, and to send it around the far side of them with 'go back' (see Figure 19.1).

The pup should quickly get the idea to go around to the far side and stay there. Repeat this in both directions a number of times. Never allow the pup to cross between you and the sheep.

You are using your position to block the pup from cutting across between you and the sheep, to send it around the far side of them, but you should begin using the correct arm signal as soon as possible. So, when you are facing the pup you point your arm in the direction you want it to go. If you want it to go anti-clockwise, you point with your left arm out to your side, and vice versa.

'Here'

When you are ready to end the lesson (remember never to allow a lesson to drag on to the point where the pup becomes sick of it) allow the sheep to get into a corner. Then position yourself between the pup and the sheep, and block any of its attempts to get around them, and endeavour to make it come 'here' to you.

When it does, take hold of the *end* of the string (which is still attached to its collar), and walk away from the sheep. Don't simply drag the pup away, but *tug* and *slack* so that it comes away more or less on its own. After only a couple of lessons it will readily come off the sheep if the string is used, without trying to get back at them all the time. Fairly soon you can dispense with taking hold of the string (but it is always on hand if required).

Introduce 'behind'

Once the pup understands 'back', you can either go out into a small paddock for its further training, or continue in the yard, or a combination of both. It depends a lot on the pup.

The next positioning command to introduce is 'behind'. You can do this after the pup has had a couple of lessons in 'go back', and basically understands to go to the far side of sheep and stay opposite you.

'Behind' tells the pup to cross behind you (see Figure 19.2). At this stage it is basically the opposite of 'go back', because the pup comes towards you instead of going away. So if it was on your right-hand side it would come around behind you out onto your left-hand side, and vice versa; therefore 'behind' can be in either direction, as can 'go back'.

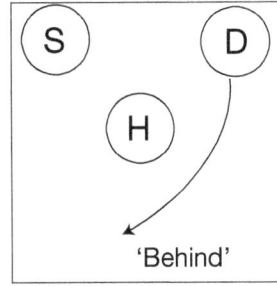

Figure 19.2 'Behind'

'Behind' is mostly used when droving a mob along, to bring the dog behind you and out onto your other side, but it is also used at other times.

To introduce 'behind' in the yard, simply step in front of the pup to block it from going around the far side of the sheep (as you have previously taught it), and wave your arm behind you and say 'behind'. Or out in the open do it on a fence. The pup will soon cotton on.

Intersperse 'behind' and 'go back' throughout the lesson.

'Sit down'

As these lessons are repeated, preferably every day if you have the time, you can begin to teach the pup to 'sit down'.

So now each time I let the pup off for a lesson with the sheep, but before going *to* the sheep, I give it a lesson in 'sit down' (as explained in Chapter 23, 'The inhibiting commands') and 'here'. Or, if it were a pup easily put off working, I would give it these lessons *after* its run on the sheep instead of before, until it gets used to them. Once the pup understands these lessons at home I then begin giving them in with the sheep, before starting each lesson in 'go back' and 'behind'.

Remember that the pup will tend to associate any lesson and command with only limited places initially. So although it may 'sit down' in your backyard, it will not do so with sheep until you repeat the lessons there. The same principle applies as you progress further with training – just because the pup obeys 'sit down' when it is *facing* you does not mean that it will obey 'sit down' when it is going away from you, such as when casting. You need to get control in each new situation.

So you can now begin getting control of 'sit down' on sheep. Start by only giving the command when you are in a position to block the pup and to ensure that it lies down. Also remember the important principle for gaining control, and never allow the pup to break the command. Tell it to 'sit down', and correct it *if* (and only *if*) it moves before the next command is given.

If you follow these principles you will be amazed at how rapidly and easily you gain control.

'Stop there' or 'sit down' arm signals

The correct arm signal for 'stop there' (standing), or 'sit down' (lying down), is to raise your hand with your palm towards the pup. You should use the correct arm – if the pup is going clockwise (round), use the opposite arm to give the stop signal, so use your left (over) arm. *This is the opposite of blocking it.* Once the pup learns the arm signals and directions later in training, it will begin to respond to this arm signal by stopping so as to change direction, but will then stop instead of continuing in the other direction.

Summary

By now the pup should be going well, either in the yard or in the open, and should basically understand 'go back', 'behind' and 'sit down', and 'here' to come away at the end of the lesson.

Mustering

(Moorlands Georgie)

Georgie brings sheep in while mustering...

...drives them across the face of the hill...

...and over a suitable crossing place.

Working a mob

(Campaspe Bounce and Cullens Kiara)

Having been mustered, a mob of 1200 or more Merino weaners is under way

Letting them poke along quietly

Putting them through a gateway

Working them through

Nearly done

Heading over the hill

Working the mob in amongst some box timber

Heading down towards the creek

Droving them along the creek bank

On the creek flat, nearing a suitable crossing

With the mob held at the crossing

Having been started across

As the sheep cross, the stockman moves in to regulate their flow

Finally, the last of the mob make their way across, and the journey can continue

20

The directional and positioning commands

The next stage is to introduce the directional commands, to be able to move the pup clockwise or anti-clockwise around the sheep.

If you were using a simple (mechanical) commanding system, with only two directional commands and no other positioning commands (no 'go back' or 'here' or 'behind'), then you would have started to train the pup to them right from the start, instead of using 'go back' or 'behind'.

But you would never end up with a well-trained, purposeful dog for practical work – acceptable maybe for mechanical-style trial work, but not for purposeful, high-quality practical work.

Introduce 'round' and 'over'

'Round' tells the dog to move around the sheep in a clockwise direction, and 'over' in an anti-clockwise direction.

With a fully trained dog, it must not change from a driving to a pulling role, or vice versa, when responding to either of these commands. Only if 'back', 'here' or 'come' is used does it then know to change roles. But we do not worry about this yet, because it would only confuse the pup, and so we require it to go right around the sheep to these commands.

The first step in teaching 'go round' and 'get over' is simple. In the pup's lesson in the yard, just replace 'go back' with the directional command. So you go in one direction, and the pup goes in the same direction to keep itself on the far side of the sheep. This is called 'counter-balance', the same as when teaching 'go back'. We also replace 'behind' with its corresponding directional command.

Never allow the pup to change directions on its own accord after you give a command, such as when the sheep move and the pup decides to cover them in the other direction. Once you have given a directional command ensure that the pup goes in that direction regardless of what the sheep do.

Early lesson overview

In these early lessons we can now incorporate all of these elements – 'go back', 'behind', 'go round', 'get over', 'sit down', and 'here' (to come right to you). It will have taken a week or two to get to this stage, depending on how often you train the pup, and on how keen it is. This will give it a basic grounding which now needs to be built upon to increase its understanding and for you to gain further control over it.

Once it has the basic idea we can take it out into a small paddock, still with quiet sheep. Out in the paddock it is a sound idea to have a light cord, five or 10 metres long (or longer if need be), attached to the pup's collar at all times during its early training. If you do this then it will not get into bad habits, and will not learn that it can get away with things in a bigger area.

Short casts

Out in the small paddock we can now introduce short casting. Give the command to 'sit down' when the pup has the sheep in a good position, and walk over and take hold of the cord some distance away from the pup. If it tries to move off you can use the cord to enforce 'sit down'. Then call the pup to you ('here') and walk away from the sheep for some distance. Command it to 'sit down', and stand between it and the sheep, facing it with your back to the sheep. Then send the pup 'go back' in the required direction with the correct arm signal.

Depending on how good the pup's natural cast and draw are, you can rapidly increase the distance of these short casts, but never so far or in such situations that it is likely to make a mess of things.

Also don't try to get it to 'sit down' at the top of the cast yet, because you probably won't have enough control to achieve this. If you give commands that won't be obeyed, the pup only learns to ignore you.

Control of 'round' and 'over'

Control of the directional commands means being able to move the dog in whichever direction we require, *without the use of any body language or position*, and, in fact, to move it in the opposite direction to that which our position might suggest.

Many handlers only train the dog using body position and body language, and then fail to gain control without these. Thus they have no control at a distance or off balance.

So it is not enough simply to have the pup go 'round' or 'over' on the opposite side of the sheep, to remain opposite the handler. More is needed. There are three exercises that I use to gain this control, and to determine if I have that control:

1   reverse counter-balance
2   stationary directions
3   overtaking directions.

Reverse counter-balance

Counter-balance is what we have taught the pup with the command 'go back'. This means the pup stays on the far side of the sheep, and if we walk around the sheep in one direction it goes around in the same direction to stay opposite us, and so counter-balances our position.

Reverse counter-balance means being able to walk around our quiet mob of sheep in one direction, and have the pup go around in the *opposite* direction. So if we walk around the mob clockwise, in the 'round' direction, the pup goes around anti-clockwise or 'over'. It is the opposite of counter-balance (see Figure 20.1).

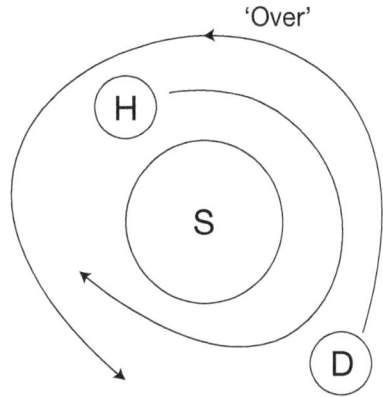

Figure 20.1 Reverse counter-balance

You can use one of several methods to achieve control over the pup to the point of being able to do this. You can work out in the open, giving the directional command and either using the long cord or 'sit down' to prevent the pup going in the wrong direction.

Or you can work in a yard such as used earlier (or even smaller if you are having trouble, such as a pen in the shearing shed) and simply use body position and voice to achieve control. This is the quickest and most effective method. But with softer dogs, or those lacking in real 'heart', it is far less effective – such dogs can tend to become slow and lack-lustre in their response.

Some softer dogs can also tend to break too wide to these commands when they are later taken out into the paddock, because they dislike them. To avoid these problems with such poorer tempered pups it is better to teach the directions out in the paddock more gradually.

Getting control of the directions – yard method

Initially, when introducing 'behind' in the yard, we stepped in front of the pup to cause it to break back behind us. We do the same thing when getting control of the directions. We start by stepping in front and causing the pup to change direction to 'over' behind us for example, the same as we did when teaching 'behind'. Once it obeys, we then follow it around a few circles (counter-balance), repeating the command any time it slows down or stops, before changing to 'round' by the same method.

Remember not to repeat the command unless you require a definite action and are going to ensure a definite response. So don't keep repeating 'over' as the pup is going 'over', when it is already doing the right thing. Because it is not having to change anything, it is simply learning to ignore you. Only repeat the command if it slows down or stops, to speed it up or start it moving again.

Gradually, instead of stepping in front of the pup to block it and change its direction behind us, we begin not getting quite so far in front of it, then back level with it, then finally behind it. We command 'over' while we are still behind it, with it going away from us in the 'round' direction. If it fails to respond by changing direction, we simply keep repeating 'over' more firmly as a correction, while keeping up with the pup and finally overtaking it to block it and send it the other way.

Immediately it obeys we stop growling at it as a reward.

As we get control of this exercise, we gradually merge into the reverse counter-balance exercise. We change the pup's direction with 'over', then continue walking in the opposite 'round' direction ourselves, and thus meet the pup head-on, whereupon we again say 'over' and cause it to continue its way around us in the same direction as it was going (see Figure 20.1).

This means we are orbiting around the sheep in one direction, and the pup in the other direction, in response to 'over'. We do the same with 'round'.

If you are having trouble, then instead of going around the sheep in the opposite direction to the pup to meet it head-on, initially you can cut through the middle of the sheep and so meet it less head-on, and endeavour to keep it going in the same direction.

Once you can get the pup orbiting in each direction freely to command, with you going in the other direction (see Figure 20.1), you are on your way to having control of the directional commands.

Stationary directions

It shouldn't be difficult once the reverse counter-balance exercise is achieved, and fixed by repetition, to now move the pup around the sheep in whichever direction we like, as far as we like, while we remain in one place. Work on this until this is accomplished, and then progress to the next exercise, which takes this a step further.

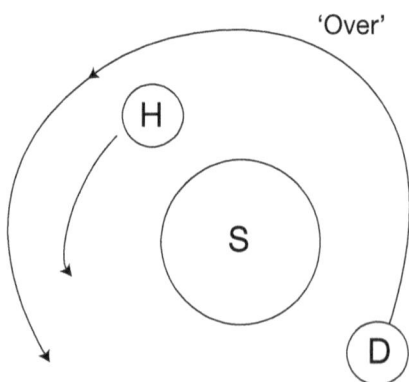

Figure 20.2 Overtaking directions

Overtaking directions

The final exercise is to be able to walk around the sheep in one direction, say anticlockwise, and have the pup go in the same direction (anticlockwise), and overtake you in response to the directional command, and so continue on around the sheep (see Figure 20.2).

This exercise can then be practised in the open paddock. Walk along with the pup bringing the sheep behind you, and have it overtake you while you keep walking in the same direction, before it goes back around behind the sheep (see Figure 20.3).

Use of commands

A reminder here may not be out of place that the way we have used the commands in these exercises is not the way they are used when the dog is fully trained. That is, when fully trained, the dog that is on our side of the mob doesn't go to the far side unless the command 'back' is used; and if it is on the far side it doesn't return to our side unless 'here' is used; and it doesn't cross behind us unless 'behind' is used.

In the early stages we exaggerate 'round' and 'over' to make it easy for the pup to understand, and then refine these commands in later stages when we teach it to drive.

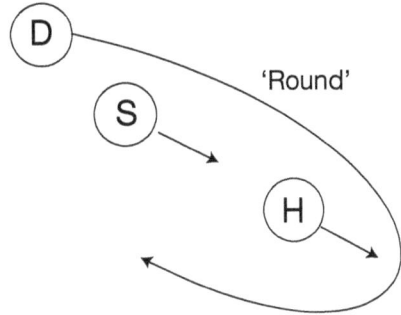

Figure 20.3 Overtaking while drawing

Getting control of arm signals and voice

These three exercises (reverse counter-balance, stationary directions, and overtaking directions) have eliminated the influence of body position, and we should now have reasonable control of the directions. These exercises should now be repeated, simply using arm signals without any spoken commands, and vice versa with spoken commands and no arm signals, so that we can control the pup by either method.

In order to get control of the arm signals on their own, we go through these exercises using the arm signal first, and then the voice to enforce the arm signal, until the pup responds to the arm signal alone. It won't take long if you have done the preceding work.

Out in the paddock mustering, at a long distance or on a windy day, you can take your hat off and wave it out to one side to direct the dog.

Introducing the lift

While this work in teaching the directional commands is going on, we can also begin teaching the pup to lift and draw its sheep properly, out in the paddock.

Assuming that we have sheep that are pretty well trained to follow us around by now, we make the pup 'sit down' on the far side of them and walk away ourselves with the sheep still following us. We then call the pup up with 'walk in' or a whistle. We can then stop the pup again and repeat the exercise, in order to teach control on the lift and draw. Gradually make the pup 'sit down' for longer periods as you (and the sheep) walk further away.

We can also introduce the pup to 'steady' by telling it to 'steady', and if it fails to slow down we command it to stop. But don't try to teach it both to 'walk in' and to 'steady' at the same time, or you will confuse it. Teach it to 'walk in' first, and only when it is confident at that introduce 'steady'.

Once again, don't give a command that you cannot enforce. So don't say 'steady' to a young dog that is rushing about, before you have enough control of 'sit down' to enforce the command, otherwise you will teach the opposite.

Summary

By now we should have the pup trained and obedient to the basic 'round' and 'over' directional commands, and to the arm signals, regardless of where we are or what we are doing. This is all the work we do on directional commands *at this stage*, but the following two exercises are undertaken later on, once the pup is out at practical work.

'Get over' or just 'over'

Once we start practical work later on, and the distance the dog must travel becomes much greater due to the bigger mobs being worked, we can refine the use of the directional commands. I do this by using 'over' or 'round' when only average movements are needed, or only 'o' or 'rou' for very fine adjustments, and then using 'GET over' or 'GO round' when a much larger or more urgent movement is required.

This is another instance where you can easily teach the opposite to what you require. The natural inclination is to use the smaller command when you think that the dog is going to go too far, and then the dog *does* go too far, and so it comes to view the small commands as meaning a big movement. Therefore you have to be very careful when using these commands, particularly in the early stages. Give the full command 'get over' when you know that the dog is wanting to do that, and the shortened command when the dog is only wanting to cover finely.

Later on you can use the commands in different situations, but make sure you enforce them. If the dog goes to make a big movement to a shortened command, stop it, and vice versa.

'Go forwards'

The final positioning command to teach is 'go forwards'. For example, when droving sheep along a road this command tells the dog to run up the wing to somewhere near the lead, to either turn the mob (maybe through a gate) or to get the lead moving and then come back the same way.

'Round' or 'over' shouldn't move the dog far enough, because when droving along a road it should stay back on the tail in response to these commands. And 'go back' would tell it to go right around in front of the lead to block the mob and turn them back to the handler. So we have a separate command to go nearly to the lead but not far enough to stop the mob. This is the last positioning command we teach, at a much later stage.

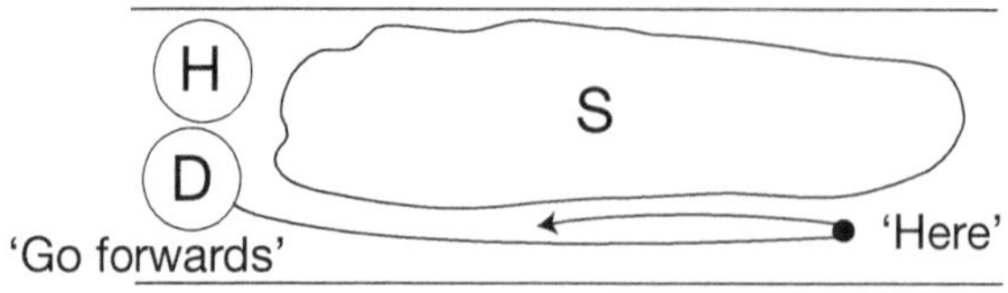

Figure 20.4 'Go forwards' on road

You can teach this during practical work by sending the dog 'go forwards' (which you shouldn't have too much trouble doing if you combine the command with a wave of your arm and some urgency in your voice) and then calling it back before it reaches the lead (see Figure 20.4). If you are consistent it should catch on quickly.

You can also do the same exercise while driving a mob along a fence line in a paddock. Send the dog up the wing on the open side of the paddock by commanding 'go forwards', and before it reaches the lead bring it back, to get the mob to string out and increase pace. You can make it 'speak' on the wing also if need be.

Also send it 'go forwards' on the other side between the sheep and the fence, to turn them at right angles away from the fence, and then drive the mob in that direction.

'Go forwards' is also useful in the yard. Instead of telling the dog to 'go back' when working the side of the race, you tell it to 'go forwards' so that it knows not to go all the way in front of the sheep to block them. The same situation applies when backing.

21

Cast, lift and draw

In the preceding work the pup has been given a good grounding in short casting, a basic understanding of the steady lift, and an idea of drawing sheep along behind us.

If it is a well-bred pup with a natural cast and draw, we will now simply be able to go out into a bigger paddock and gradually (or not so gradually with the good pup) send it longer distances. If it is well-bred it will cast out wide and deep for long distances naturally, and then drive the sheep back to us without a word of command. However, most pups will present much greater problems.

But even with a good natural pup, for mustering work in rough, hilly country, we still need to gain control to break it out on the cast or bring it in on the cast as required, and to stop it on the cast or draw at a long distance.

Lengthening the cast

There are a number of ways to quickly teach a pup to cast distance. One good way is to use a fence initially, and teach the pup to cast up the outside of the fence to find a gateway or hole to come in behind the sheep. Or send it around a big dam, or a patch of rushes, or the shearing shed.

But as it gets the idea, you can also use the lie of the land to your advantage, by where you position the sheep, and where you send the pup. For example, you can use a gully in the right place, because a dog casting in hilly country will sometimes follow a gully, particularly if casting uphill, because the gully seems to be the easiest route.

Also if you cast the dog across the side of a hill, so that it casts downhill below the sheep and must then come uphill to get behind them, this downhill run will tend to keep it out wide. Conversely, if you cast it above sheep on the side of a hill it will tend to cast more narrowly (though not if it is a good dog). Another trick is to cast it while on the horse, and keep pace with it to prevent it from crossing.

These tricks help to prevent the pup from crossing, and so you can quickly teach it to cast *distance*, but they don't really teach it to cast wide and deep when those props are not present.

For example, a method sometimes recommended is to erect wire wings to keep a dog out. But the dog doesn't *keep itself out*, rather it is forced to stay out, and when you remove the wire wings it will cast narrow again. Casting a dog outside a fence can teach it the general idea of casting distance, but it will not teach it to cast properly. You must get the dog to keep itself out.

Therefore, later on you need to do the exact opposite, and cast the dog in situations where it is *likely* to cross, and teach it not to.

Widening the cast

If you have a pup that casts straight or crosses, and maybe then splits the sheep up the top, or rings around them, then you are going to need to get it to widen out (if you decide to persevere with it).

The common method is to simply try to chase it out, or have someone else near the sheep to chase it out. This is not much of a method, although it can be very effective if you are mounted on a handy pony. But if you want a mustering dog then sooner or later you have to teach it to break out on its own, and I prefer to do that right from the start.

Teaching 'get out' and 'come in'

For a poor casting pup I recommend using a dam to break it around, because the size of the dam teaches it to break out of contact with the sheep. For a good pup I use a big tree, because the pup's *breeding* tells it to break out of contact (the tree just gets it started), but a dam is good also.

Remember, however, that a poor pup will never be much good in difficult situations, no matter how much training you put into it.

We start off by teaching the pup the basic idea of casting around the dam. We position the sheep near the dam, walk away from them around the dam about quarter of a turn, and cast the pup around the dam away from them (see Figure 21.1).

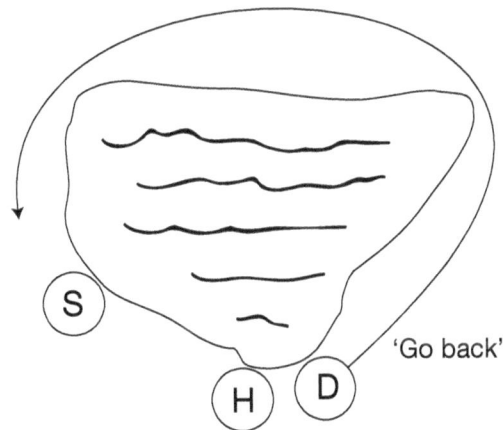

Figure 21.1 Introducing pup to casting around a dam

Once it has the idea of casting around the dam we begin to position the sheep and ourselves further from the dam, but not too quickly, little by little. We have the pup on a

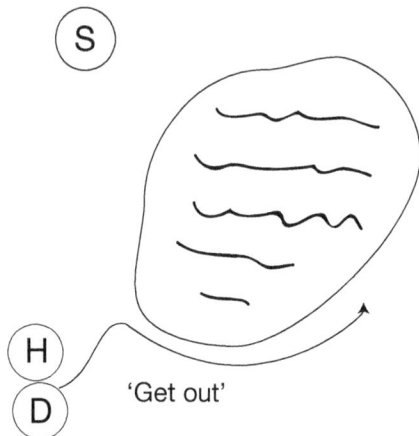

Figure 21.2 'Get out' around dam

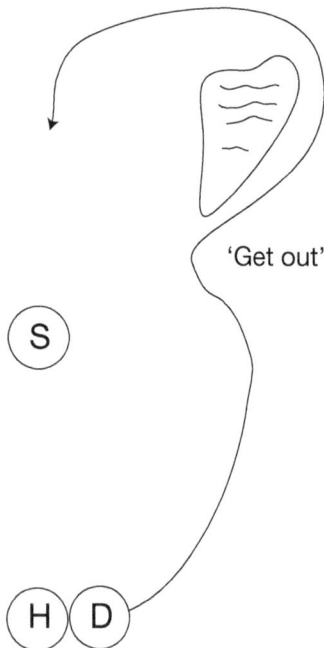

Figure 21.3 Deepen cast

long cord, and send it to cast around the dam with 'get out', from maybe 10 metres from the dam. If it attempts to cut in, the cord prevents it. Then we try again.

Because cutting in never allows it to get at the sheep (it is prevented by the cord), it will soon learn to break out around the dam (see Figure 21.2).

We then move further and further back from the dam, until the pup will cast around the dam as required. Once we reach this stage we send it out with 'go back' instead of 'get out', and then when it gets level with the dam we stop it ('sit down'), and tell it to 'get out' to break out around the dam.

By the position of the sheep and the dam, we can teach it to break out to *widen* the cast, or to *deepen* it (see Figure 21.3).

Now if the pup has any brains it will begin casting around the dam on its own initiative, without being stopped or broken out. But we still don't really have *control* of 'get out' (we have taught it to cast around the dam, not to 'get out'). So once it reaches the stage of breaking out around the dam without being told, it is time to teach it to 'come in' on the cast.

So we give it a few more runs breaking out to let it know that this is right, then next time we call the pup to 'come in', to cast around the sheep *inside* the dam, using 'sit down' or the rope to ensure that it obeys.

This 'come in' command is valuable in practical work, for example if you have a mob of sheep or cattle that are racing to box (join) with another mob. If you have a handy dog and send it after them, it will break out and go around the far lot instead, to muster the whole lot. But if you have this 'come in' command you can call it in between the two lots and prevent the box up, particularly if you have it trained to split sheep (see Chapter 28). It is also useful in general work to tell the dog where sheep are to be found, because you have a much better view than the dog does from ground level.

We can then begin alternating between the two, sometimes breaking it out and sometimes pulling it in. *It is this alternation between the two opposing actions that really teaches the pup the meaning.*

We are starting to get control.

We should now work on extending the distance we cast it, and phase out the stop before we break it out, just breaking it out with 'get out'. Only stop it if it fails to respond. The next phase is to try something less obvious than the dam, such as a tree.

By use of control, or the long rope, you should now be able to break the pup out with nothing for it to break out around.

Another method

Another way of teaching a pup to break out is simply with the sheep in the middle of a paddock. It is particularly suited to dogs of UK blood, which have been bred for their ability to be broken out on the cast (and to 'square flank' to command). They tend to look out very readily, and follow their head (what I mean by this will become evident as you read on).

You stop the pup near the sheep, and stand to one side of it, and tell it to 'get out' (see Figure 21.4). Then when it comes back in again you stop it, and walk over to it and repeat the exercise, and repeat this over a few days, until it begins to anticipate. When you see it begin to *anticipate* and *look out* while it is lying down (in the position as in Figure 21.4), say 'get out' to reward this looking out, and repeat this exercise until it gets the idea, waiting for it to look out before you give the command.

Then you cast the pup from further away, with the long cord on it. Stop it before it gets level with the sheep, and walk up in front of it and have it 'get out' (see Figure 21.5).

Do this a few times, and watch the pup. It should begin to anticipate, and when you sit it down will look out at right angles in anticipation of the command. When it does so, give the 'get out' command as a reward. Give the command without getting right in front of the pup, when you see it look out. Don't let it move off without being told though, otherwise your 'sit down' command on the cast will come to mean 'get out'.

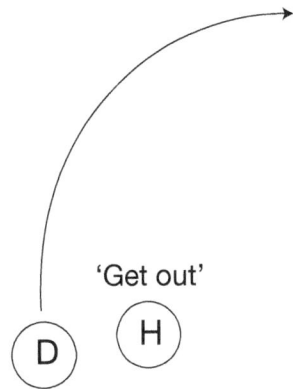

Figure 21.4 Another method

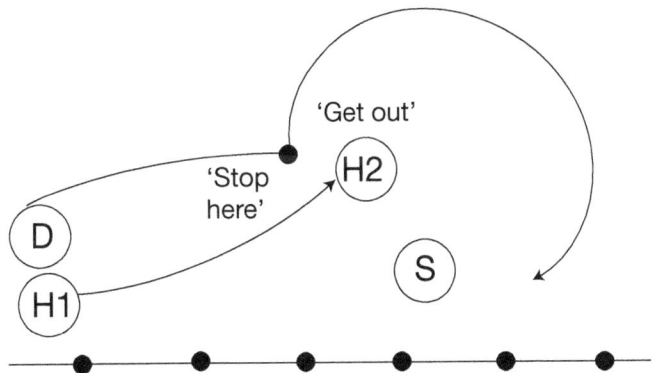

Figure 21.5 Another method – stage two

This was how I taught Boho Misty to break out on the cast. As she was exceptionally intelligent, I only had to break her around the sheep a couple of times with the command 'get out', and the next time as I walked up towards her to stand in front of her to break her out, she turned her head and looked out in the direction I was going to send her, anticipating it. So before I got there I said 'get out', and she did.

I then cast her from a longer distance, and stopped her, and *waited until she looked out of the cast*, then immediately said 'get out'. Because she had suitable instincts, and great intelligence, it took about two lessons to teach her that, and she never forgot it.

This method works well with some dogs, but other dogs will tend to simply cut in again as soon as they get around you (see Figure 21.6), and don't head out in the direction they are looking, and often don't have the intelligence to figure it out so quickly. They will take much longer to teach this.

Also this method (of giving the 'get out' command when the dog is looking in the direction we want it to go) is valuable for the mustering dog. When out mustering we can stop the dog on the cast, and it has learnt to begin looking around, so we wait until it is looking in the right direction and then send it out. Thus we have a good method of telling it in which direction to go.

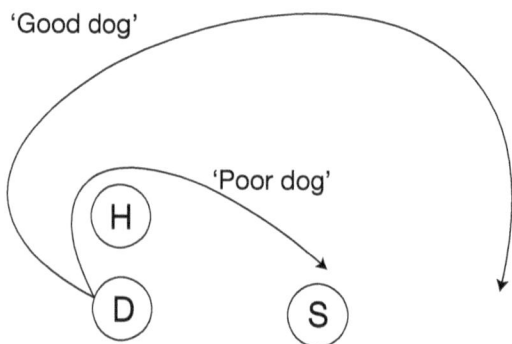

Figure 21.6 Good dogs and poor dogs

This is the best system for practical mustering work in rough hill country that I know, but it only works well with some dogs. We can break the dog out 'on the run' most times, but in really tricky conditions we can sit it down and let it look around until it looks in the right direction. In this way you should be able to direct it to any place you like.

Yet another method

With some pups, particularly very excitable pups lacking much ability, they will still tend to cut in straight towards the sheep and sool around them, even after they break out around the dam or a tree. For this type of pup the best bet (if you bother with it) is to teach it 'out of it' to move directly away from the sheep (on a running wire as explained in Chapter 23), and then make it cast the fence in a small paddock.

Stand between the sheep and the pup with the sheep in the middle of the paddock, and with the pup stopped halfway to the fence. Make it turn around and 'out of it' to the fence, and then make it go 'round' or 'over' around the paddock hugging the fence.

Walk around the sheep in the middle of the paddock, keeping between them and the pup, and any time it comes in make it turn around and 'out of it' back onto the fence (see Figure 21.7). You will soon have it casting the fence in this manner (unless it lacks heart, and quits). This 'out of it' command is also the best way to teach most dogs to work wide (and to teach them some 'distance'), far more effective than chasing them out by throwing things.

You can then graduate out into a bigger paddock, and working off a horse is best. You can ride inside the pup's cast and command it 'out of it' any time it comes in, and force it to cast wide and deep. Of course such a cast is completely man-made, and the dog will be hopeless in tricky situations left to its own devices. But with poor dogs you have no choice (other than to get a better one).

Further work on 'get out'

Once the pup understands and obeys 'get out' reasonably, there are a couple of useful exercises to reward it for breaking out and to encourage it to enjoy the exercise, instead of viewing it as a chore.

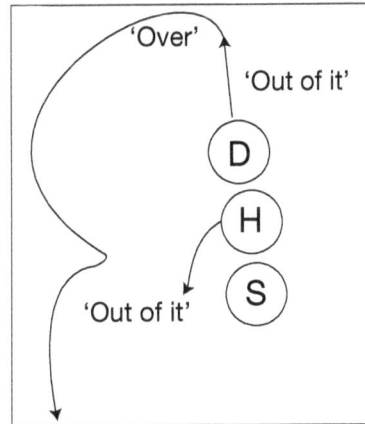

Figure 21.7 Cast the fence, or work wide, with 'out of it'

Place some sheep in a position such as in a corner of a paddock over a rise out of sight, and cast the pup in such a way that it will probably miss them. Then use 'get out' to put the pup out around them. In this way it will learn that it pays to obey 'get out'. This will greatly assist in gaining *willing* obedience, which is the only sort if you want a high level of work. Do a similar thing with the 'come in' command.

Casting the dog blind, with the sheep in a different place each time, is ideal for teaching it to listen to your commands, and also to look for sheep (but looking for sheep is largely an inherited ability).

Slingshot casting, and positioning the dog at the start of the cast

A method often advocated to teach a dog to cast wider is to start it on the opposite side of you from where you want it to cast, and then bring it around behind you, so that it tends to curve around you to begin the cast. This method is sometimes referred to as 'slingshot casting'.

Some people also advocate casting the dog from a long way behind you as another ploy in getting it to start the cast wider. Either of these methods has its place when first getting a pup started casting, *but you must remove any reliance on them if you want to teach it to cast properly.*

If you can't sit a dog directly in front of you, between you and the sheep, and have it make a perfect cast from there, then it isn't trained (or bred) to cast properly. When mustering we often recast a dog when it is hundreds of metres away from us, and a slingshot cast or other ploy is of no use then; the dog must be trained (and bred) well.

Handler's position

Most Australian handlers stop their dog and stand between it and the sheep. They *face the dog* with their back to the sheep, and cast it out from that position. Once again, this is useful for a young pup that you need to keep your eye on, to prevent it from going before it is told, but it shouldn't be required for a trained dog. You should have it well enough trained that it will wait for your command.

The big problem with always facing the dog to cast it is that you don't give it any clues as to where the sheep are.

It is all right on a small trial ground or in smallish paddocks, where the dog knows where to find the sheep. But in practical work it is far from satisfactory, because many times you need to recast and redirect a dog from positions hundreds of metres away. In such situations when mustering in rough country, or in any blind casting, any clue that you can give the dog is valuable. The main clue is to face your horse, or yourself, in the direction that you want the dog to cast. That is, you point your horse towards the sheep, or even move in that direction (this is another reason why 'chase-'em-round' arm signals are not ideal).

So once your pup has the idea of casting, and is reliable and remains sitting down awaiting the signal to cast, you should begin casting it with *your back to it*, facing the direction of the sheep. Then, whenever you need to give it more commands on the cast to help it find the sheep, always face in the direction that the sheep are to be found, or walk in their direction.

The only time you should face the dog is to recast it directly away from you with 'get away' to go for sheep *behind* the dog. Once again you would be facing the direction of the sheep, because they would be behind the dog.

Therefore the handler who always faces the dog to cast it (and most in this country do) has a problem: sometimes they are facing away from the direction of the sheep (when they usually cast it 'go back') and sometimes they are facing towards the sheep (when they recast it 'get away'), so there is an inconsistency and the dog cannot take its direction from the handler.

Teaching the dog to look for sheep (or cattle)

You can teach a pup to look for its sheep before it is cast out, although many times it will be cast out blind in practical work. This is simple to achieve by using the casting command 'go back' as a reward command.

Set the pup and yourself up ready to cast, and watch its expression. Tell it something like 'sheep', and wait until you see that it is looking at the sheep (start with them reasonably close by). While it is looking at them, give the command to cast – 'go back' – to reward this behaviour.

If you are consistent it will begin looking for its sheep in response to 'sheep'. (Personally I never bother because I like dogs that are bred to look for sheep on the cast, and they are often cast off blind anyhow. A dog should just cast, and cast, and cast, searching for stock, but most dogs won't.)

You can also teach the difference between sheep and cattle in this way. Teach the dog the command 'cows' to look for cattle before casting it out. Then whenever you are sending it out into a paddock with both sheep and cattle use the relevant word, and wait until it is looking at the right mob, and send it on its way.

Lift and draw

The main requirement for the lift and the draw is that we have control to be able to stop the pup up the top of the cast, and to stop it or steady it on the lift.

The quickest way to get this control is with the long cord (although with a good, calm pup devoid of busyness, and with a natural lift, you won't need the cord). Cast the pup out with a *long* light cord on it, and when it gets near the end command it to 'sit down'. If it fails to respond, use the rope to ensure that it does, and then send it on to finish the cast.

However, I find that it is important to teach a whistle for 'sit down', because the response at a distance is better. This whistle is a single monotone whistle that can be held for an extended length of time. The fact that it can be held in this manner is one of the reasons why it is more effective than a vocal command – you can create a 'negative reinforcement' situation where you hold the whistle until the pup responds.

The other reason why whistles are more effective is because they retain their same sound at any distance, whereas a shouted command sounds different to a softly spoken command, and so is really a different command as far as the dog is concerned.

So the first thing I do is to go back and teach the pup to lie down in the backyard to this whistle. Then I introduce it while working sheep close to me, and then further away. Next I introduce it on the cast with the long rope. You should soon have control of the pup up the top of the cast. If you stop it every time up at the top then this will become a habit, and it will be much more likely to obey.

Another method is to work the pup from a horse. Cast it out and then ride up to be near the sheep when it gets there, to make it stop. With repetition it will get the idea, and you will be able to stay further and further back from the sheep.

Some people use the crack of the stockwhip to stop the dog at a distance, but I don't like this because I want the dog to force when the whip is cracked, and not lie down. However, the noise of a whip can be effective in freeing up a sticky dog when casting – it works with some dogs, not with others.

Selection of stockwhips – red hide, kangaroo hide and white hide

With a pup that rushes its sheep on the draw, it is an advantage to teach it to 'out of it', that is, to turn around and run away from the sheep (see Chapter 23). You can then cast it out, short distances at first (a narrow laneway is good), and when it rushes in behind the sheep reprimand it with 'out of it', and make it clear out. This works better than simply stopping it.

Also, when the pup is stopped it cannot cover its sheep, but when you teach it 'out of it' you can also teach it to work wide, and prevent it from chopping in if need be.

Directional commands on the draw

To get control of the directional commands on the draw, two commands are of great help. Control to stop the pup is vital, and with some pups 'out of it' is valuable to turn it around and make it go away from the sheep.

If you give a directional command and the pup fails to obey, stop it and repeat the command until it obeys. Also try sending it 'out of it' and not letting it come in towards the sheep until it responds to the directional command.

However, teaching the pup to drive sheep away, and to cross and ring sheep on the draw to command (see Chapter 24), is invaluable to help get control of the directional commands on the draw.

The dog only trained to pull, and not trained to drive away, will always tend to resist the directional commands on the draw, because it has been taught to always stay on the other side of the sheep.

But if you have it trained to drive away, and to ring around to the head of the sheep on the draw, it will be much more willing to move freely in response to the directional commands.

Summary

By this stage we should have a young dog that will cast long distances freely, wide and deep behind the stock, and break out or come in as required.

It should also lift and draw them steadily down to us, and obey the 'sit down' whistle on the cast and the draw.

22

Heel

It is of great value – to the practical stockman – to teach the dog to heel, that is, to follow *behind* the stockman. This contrasts with the obedience school method or the common UK method of heeling up *alongside* the handler. Positioning the dog behind you means it is less likely to creep in front of you, and it can be sent out to either side as required.

Once you teach a dog to heel properly it will tend to fall into this place even though no command is given, instead of rushing about and having to be called back all the time. And the same applies when you are tailing a weak mob along in the paddock or on the road with a pushy dog; otherwise you are continually calling 'here' to have the dog return to your side, before it is drawn back towards the stock. A few days of this can become very irritating and tiring. It is much better to have a good 'heel'.

'Heel'

I like to have the pup going fairly well before teaching it to heel. This is because with the pup running around freely there are more opportunities for teaching 'here' than if it is tagging along behind all the time.

The initial training of 'heel' is very simple. Put the pup on a lead and equip yourself with a leafy branch (your hat is also a good option because you always have it with you). As you walk along, swing the branch or your hat like a pendulum from side to side to block the pup from passing you. It will quickly get the idea.

Each time it tries to get past and you block it, give the command 'heel'. Don't repeat 'heel' while it is following well, but only just before you make it go back to the desired position, because that is what you want 'heel' to mean.

Gradually get more exacting. Start walking in circles and figures of eight, in and out and around trees and the like, and finally turning on the spot. This last exercise of turning on the spot can also be done up against a fence to good effect.

As in all training, repetition is a great tool. If out doing a day's fencing it is a simple matter to attach a lead to the back of your belt for half-an-hour or so, and the pup begins to feel at home at your heels. You can do the same thing while taking the dogs for a walk, or while feeding the other dogs, and so on.

The pup has now learnt to follow behind you. The next stage is gaining control, which can be more difficult in trying situations. A couple of principles discussed previously can be applied here:

- don't anticipate disobedience, but only correct it; and
- at times allow the dog to think that it has got away with disobedience before correcting it. This will keep it guessing.

So, stop waving your hat to keep it behind; instead, wait until it gets in front of you and then use it. And then stop using a short lead – use a long lead so that you still have ultimate control, but don't use it to keep the pup in place. This is how you graduate to off-lead work – the long lead doesn't *prevent* the pup from breaking 'heel', but it does allow you to correct it if it does, and to give it no option other than to return to your heels.

You also need to teach the pup to heel behind the horse, which will only take a couple of lessons once the preliminary work is done. If you have used your hat to block the pup, and if it fails to respond to 'heel' when you are on the horse, you can often just lift your hat off and the pup will come back to 'heel'.

The final icing on the cake is to make sure that the dog will heel while you are working a second dog, but only dogs with good, calm temperaments will ever be much good at this.

23

Inhibiting and wide-working commands

As mentioned elsewhere, 'sit down' is the foundation of real control. You could use 'stop there', to stop the dog on its feet, if you wanted to, but with most dogs it is more difficult to get control of than 'sit down', because they can 'creep'. 'Stop there' standing is more useful for trial work, and for really getting a dog to obey 'steady'. For practical work 'sit down' is better, as the foundation of control.

Also it can be better to have the stop whistle for 'sit down' rather than for 'stop there'. This is because it can be easier to get control of the stop at the top of the cast, and on the draw, if you make the dog lie down. I use a whistle for 'sit down', and one for the lesser 'stop'.

One reason many handlers lack real control of their stop command is because they only have one command with which they attempt to do three things – it is used for 'stop there' (standing), 'stop' (pause) and 'sit down' (lie down) – which can never work.

The other inhibiting commands to teach are this lesser 'stop' command, 'steady', 'stay', and the wide-working commands 'keep off' and 'out of it'.

'Sit down' (lie down)

Teaching 'sit down' to make the pup lie down is very quick and easy. Have the pup on a lead and push down on its withers until it is straining against you to hold itself up. Maintain this position until it eventually lies down.

The important point is that the pup is lying itself down, not you pushing it down. If you can follow this principle in any training situation then the pup will learn much more rapidly.

The common method of teaching this is to push the pup all the way to the ground, while commanding 'sit down'. This works, but inefficiently. You should only push it part way down, and then let it put itself down the rest of the way. Another similar method is to put a lead on the pup, and run this under your boot, and apply pressure so that the lead is holding the pup halfway down, and then wait for the pup to lie itself down.

Then begin extending the time that you require it to stay lying down. Begin walking around behind it and stepping over it, and walking further away. Eventually you can hide in a shed or something, and observe it through a nail hole.

Never let the pup get up until it is told to.

A good extension to this is to have the pup remain lying in one place while you have the other dogs off and are feeding them or something. This is also a good introduction to working more than one dog together.

Another exercise is to have the pup remain lying down for half-an-hour or so while you read a book – have the long rope on it. Remember to only give one command, and don't repeat the command unless the pup moves. This extended 'sit down' is something that most people neglect, and so they have problems getting a dog to remain in one place for any length of time. It is always expecting to be moved soon, and has no concept of remaining in one place for any length of time.

Also when teaching 'sit down' you can give the pup some idea of 'here' and 'walk in'. For example, after it has been sitting down, you can sometimes have it come *all the way* to you with 'here'. Don't sit it down halfway or its response to 'here' will diminish. Instead, at other times call it towards you with 'walk in' (in place of 'here'), and then 'sit down' when it is part way towards you. This also introduces the steady lift.

'Stop there' (standing)

Teaching 'stop there' is very simple. Just put the pup on a lead and walk backwards away from it, and block it from coming towards you with a stick, commanding it to 'stop there'. Don't let it move until you call it 'here'. However, don't repeat 'stop there' *unless* it moves; that is, don't repeat the command as a means of *preventing* it from moving, but only as a *correction* if it does.

Some pups, however, may lie down naturally, either because they are soft in temperament or because you may have taught it to 'sit down' first, and now you want to teach it to stop on its feet. In these cases, an exercise in teaching the pup to sit down and then stand up is valuable. I find the best way is first to teach the pup to walk backwards, as explained below ('keep off'). Then you can use the lead to draw it forwards, and the command 'keep off' to cause it to walk backwards, and so keep it on its feet or make it stand up. By judicious use of these, and the command 'stop there', it will soon understand to remain standing.

(Contrary to the opinion of some Kelpie people, I don't believe the Australian Border Collie is any more naturally inclined to 'clap' (lie down) than the Kelpie. Most borders are taught to lie down rather than doing it naturally.)

'Keep off' (walk backwards)

Teaching the pup to walk backwards has some benefits, even though a good judge (and they are few and far between) rightly penalises a dog for walking backwards in a sheep dog trial in most situations. This is because many handlers use it as a ploy to get sheep to turn and walk off a weak dog – they use it to 'drift' sheep, rather than getting a 'grip' of them and putting them around the course.

One use for 'keep off' is as a correction for a dog that doesn't obey 'stop there' (standing) quickly enough, or creeps forwards. You can send it back to the position that it should have stopped in, and so not to allow it to benefit from disobedience. Another use is as a correction for not obeying 'steady'. And, as mentioned above, it is useful to keep a dog up on its feet when teaching it to 'stop there'.

Also 'keep off' can be useful when teaching a weak pup to walk in to force, because you can alternate between 'walk in' and 'keep off'.

To teach 'keep off', stand close to a fence and face parallel along it, with the dog on a lead in front of you facing you. Hold the lead in your left hand (assuming the fence is on your left) and your hat or a leafy branch in your right hand. Command the pup to 'keep off' while you walk towards it. The lead prevents it from turning its back to you, and the fence and the hat prevent it from slipping around to either side.

Then walk backwards away from it and command it to 'walk in' towards you, before changing back to 'keep off'. It will quickly get the idea. You can also do something similar in the handling race in the sheep yards.

You then transfer this technique to working with sheep. Position the sheep in a corner with you standing between the sheep and the pup, and make it 'keep off'.

'Stop' (half-stop)

The 'stop' command differs from 'stop there', in that the dog must stop moving but is then allowed to move off on its own accord, and to continue covering the sheep. It is most useful when working touchy sheep such as in a sheep dog trial, but often in practical work also.

As the dog runs around you can stop it where you wish, but it is allowed to keep covering its sheep from that position, and so continue to hold them. When working a mob you might send the dog up one wing, and tell it to 'stop' where you want it, but then it knows that it is allowed to move and cover or force on that point.

A good method to teach 'stop' is to stand between the pup and the sheep, facing the pup, and tell it to go 'round' five metres, then 'stop' and immediately 'over' back past you the other way, then 'stop', 'round' again. Give the next directional command before the pup has time to lie down, any time it looks like doing so.

The other good time to teach 'stop' is when teaching the pup to drive, as explained in the next chapter.

'Steady'

The simplest way to teach 'steady' is when the pup is drawing sheep along behind you. Tell it to 'steady', and then make it 'stop there' or 'sit down' if it doesn't ('stop there' is better, because it keeps the pup on its feet). Then call it in again and repeat this, until it understands that if it doesn't slow down it will be stopped. It learns that slowing down is preferable.

Having taught the pup 'steady' with it coming *towards* you, however, does not mean that it will obey when going around sheep, or when casting. But you can use the same technique to get it to go slow on the cast. This works well with a natural caster, but tends to cause a made caster to come in narrow.

At times it is desirable to be able to slow the dog right down up the top of its cast, such as when casting around some cattle that are in such a position that a fast cast might spook them and cause them to jump a fence. However, a good dog casts at a sensible pace, and thinks and looks while it is casting, and *steadies itself* on the lift.

Another method is to use a long rope to steady the cast. The advantage of this method is that you don't rely on 'stop there', and so don't get the pup sick of that command. Cast the pup out, and follow it yourself with the long cord on it, and in this way you can cause it to cast at a walk in response to 'steady', if you want to.

Another idea is to put a fair number of sheep in a yard, so that the pup is moving more or less directly away from you as it goes around them. Have it on the rope, and teach it to go around them 'round' or 'over' away from you, steadily. If you haven't got enough sheep to do this, then you can put two or three sheep in a small pen standing in the open, and work the pup *outside* the pen, on the rope.

These extra exercises in 'steady' are really only applicable to trial dogs.

'Stay'

'Stay' is the command that I use to tell the dog to remain in the back of a ute or something of that nature, but that it is free to move around in there.

I begin by teaching the pup in its dog yard, by opening the gate and telling it to 'stay', and not allowing it to come out until called. You can teach the pup this before any work on 'stop there' begins, and in fact you can teach whole litters running together in a yard, three or four months old, at the same time.

'Out of it'

'Out of it', taught in the following manner, is a very useful command in various situations. It is the best way to teach a dog to work wide, far better than throwing lumps of dirt at it and so on (although that can work with very soft dogs, but isn't particularly successful with strong, hard workers). It is also useful as a correction for not obeying 'sit down' or the directional commands on the draw, or rushing sheep on the draw.

Also it is the ideal command when counting sheep through a gateway, or when wanting a dog to ease the pressure off sheep in any way.

The best way to teach 'out of it' is to attach the dog to a running wire by a short lead. This running wire is simply a length of fencing wire strained between two pegs along the ground. It keeps the dog going in a straight line away from you, even though there is no fence or anything to guide it. Indeed it is *better* out in the open with no fence nearby. This is why it is better than teaching it in a handling race or somewhere like that, because the dog learns to go straight without anything to guide it.

Once the dog understands to travel the length of the wire away from you in response to 'out of it', you then graduate to working with sheep in a yard, then out in the open (a narrow laneway is a good first step).

Summary

In order to maintain the highest level of control, always try to use the least restrictive command possible in any given situation. The dog will resent the intrusion on its work least, and so better control will be maintained.

So, depending on the situation, try to use 'steady' or 'stop' in preference to 'stop there' or 'sit down'. This is a general rule, and like most rules there are exceptions.

Try to use no commands at all whenever possible.

24

Driving

There are many advantages to teaching a dog to drive stock *with* you (as against pulling them *to* you), and particularly in teaching it to drive stock directly *away* from you.

Most stock work involves the dog driving stock *with* the handler, so that basic level is essential. But more than that, when mustering in rough country you often require a dog to take sheep or cattle in a direction other than straight towards you. This might be in order to put them over a creek crossing or around an eroded gully, or along a fence line to a gate, or to join up with other stock, or to take stock across a particularly rough bit of country that you would rather not go yourself, and so teaching to drive away is essential.

Teaching to drive away can also result (with a very good type of dog) in helping it quickly become a very purposeful, brainy worker, where it develops enough sense to drive sheep to the gateway or wherever they are required, without constant commanding, instead of always bringing them to the handler. Not many dogs have the ability or temperament or brains to do so, but a few do, but they must be taught to drive away.

Also, when you can direct the dog that is driving sheep away it means that you can also prevent it from crossing the cast, in conjunction with 'get out', or cause it to deliberately cross the cast as is sometimes needed.

And, if you don't teach a dog to drive away from you (crossing between), but only to drive with you (crossing behind), then it often tends to slide away from you while droving a mob, and not to drive close beside you when that is required. Teaching it to drive away directly in front of you minimises this problem, even when later driving and crossing behind.

Also if you don't teach the dog to drive away, and to cross to command, you will never get real control of the directions on the draw.

Other than these aspects, teaching to drive will also result in much greater control, and make the dog more pliable.

However, in most situations when droving stock, whether that be along a road or in the paddock, I prefer the dogs to cross behind me. In that way I can regulate the pressure being applied to the mob – if I back off to ease the pressure, so do the dogs, which generally fall in at my heels. If I come in closer to the stock, the dogs begin working again. They should only cross and drive away when told to do so.

When to teach to drive

I like to have a pup casting 200–300 metres well, and drawing sheep to me well from that distance before I start teaching it to drive. I also want good control of all the other directional commands and 'sit down'. This is usually at about nine months of age, depending on how good the pup is.

Driving with the handler

Teaching to drive with the handler (crossing behind) is fairly simple. Using the commands 'walk in' and 'stop' and 'to me', and 'over' and 'round' and 'behind' (see Figure 24.1), you shouldn't have much trouble with your quiet sheep, and breeding and experience are the main requirements.

At this stage all we want is the pup to follow the sheep in response to 'walk in', making no attempt to head them. So we position the pup close to one side of us, take hold of the cord, and ask it to 'walk in'. *The long lead will expedite things greatly, and get superior results.*

Most likely the pup will begin to move around the sheep to head them, but the cord will prevent that, perhaps using the pup's name, or 'stop', as a correction, followed again by 'walk in'. If the pup moves too far away from us, we bring it back towards us with 'to me' (see Figure 24.1). (If you use 'here' instead of 'to me', your control of 'here' when you want the dog to come *all the way to you* will diminish, so I use 'to me' instead. Also a directional command would cause it to arc out around behind us, and so it would break off the sheep more, so I use 'to me' to come directly towards me.)

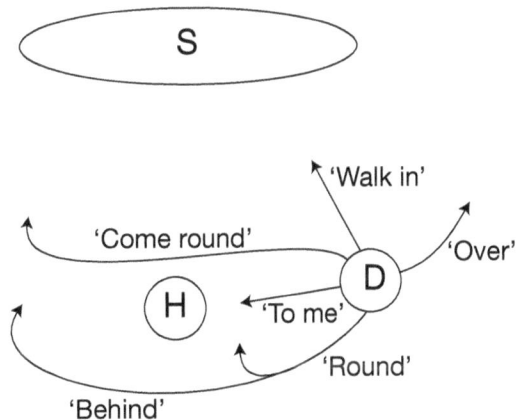

Figure 24.1 Commands when driving

This is where you will see if the pup has the natural, direct, walk-in force on the tail of sheep. The good pup will readily understand to walk in directly towards the tail of the sheep in this situation, and will get the hang of this exercise in about 10 minutes. Many Australian dogs have no idea, and won't walk in to force on the tail, and when you stop them from going around to the head they will simply stand in one place.

Occasionally you can cross the pup behind you with 'behind', and then 'stop' it and have it 'walk in' on the other side. With the cord on the pup it will rapidly understand. The cord is much better than making it sit down all the time, because it would get sick of that and its response to 'sit down' would worsen, and it would also take much longer to teach.

As these lessons progress we can also send the pup away from us with a directional command, and use the lesser 'stop' command and the cord to prevent it going too far.

In this way we are refining the use of the directional commands to their final form, in that the pup is no longer allowed to head the mob in response to a directional command alone.

Driving across, and away from, the handler

Now that the pup has the basic idea of working on our side of the sheep, we can teach it the crossing commands, and to drive sheep directly away from us. The first thing to teach is the crossing commands 'come round' and 'come over'. These bring it *between* the sheep and us, instead of *behind* us.

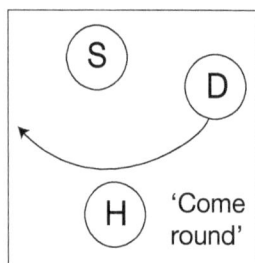

Figure 24.2 Teaching the crossing commands

Initially start teaching these in a yard. Stand with your back against the fence, and call the pup to 'come round' or 'come over' between you and the sheep (see Figure 24.2). Most likely it won't understand what you mean, and you may need a lead on it the first couple of times until it realises that it is permitted to cross.

To use the arm signals, you cross your arm in front of you. So if you are facing the sheep and the dog is on your right, call it 'come round' to cross between you and the sheep, and point your right arm across in front of you, 45 degrees to your left.

Once it crosses between you and the sheep, allow it to go right around to the far side. Otherwise if you stop it straight away it will be unsure if it did the right thing in crossing between.

Repeat this exercise in each direction until it understands. Then try to get it orbiting right around in both directions, crossing between (this is also the first step in teaching 'draw ringing', described next).

Then in later lessons begin moving out from the fence and again ensuring that it crosses between. Once it has a good grounding in this, you randomly alternate between crossing *behind* you, 'round' or 'over', or crossing *in front*, 'come round' or 'come over', to give it a clear idea of the distinction between the two.

Now we can go back to our driving lessons in the open, and instead of having it cross behind us we have it cross in front of us, and gradually have it 'stop' and 'walk in' when it is more directly between the sheep and us instead of out to one side.

Also we must get it responding to the directional commands properly when it is between the sheep and us, so that it breaks squarely around them and doesn't shoulder and cut close (see Figure 24.3).

If the pup is well bred not many problems will be encountered. Otherwise, there are a number of things you can try. One is to use the long lead to prevent it shouldering, and to encourage it to break square. Or, every time it cuts in, sit it down and wait for it to look out, then give the command to make it break square. But such a dog's performance will always revert to instinct *when the pressure is on*.

So we now gradually extend the pup's capabilities until it will drive sheep away from us wherever we wish.

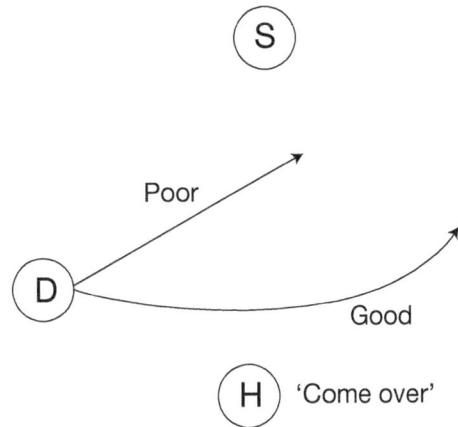

Figure 24.3 Breaking square, or shouldering, to crossing commands

Draw ringing

Another useful exercise I call 'draw ringing'. You cast the pup around sheep, and allow it to pull them *part way* towards you. Then call it around in front of them, with 'come round' or 'come over', to block them and then drive them away from you.

This helps to increase the control you have over it, and improves its response to the directional commands on the draw greatly. It also helps to make it a more thinking dog, instead of one that always does the same thing every time it is sent out, just out of habit (casts around sheep and brings them back to you). When mustering in rough country it all becomes very useful.

For example, if you cast the dog out around a few sheep in the hills, the lead sheep can sometimes come away from the dog fairly fast, and leave a crook (or cunning) sheep or two behind. If you can call the dog to come around in front of the lead sheep and turn them back to the tail, then try again, they will usually come down more steadily, and with luck the tail can keep up.

The easiest way to introduce the pup to 'draw ringing' is to cast it out around some sheep, and then climb over a fence yourself and walk back away from the sheep. The pup will draw the sheep towards you, but the fence will prevent it from getting them all the way to you. Then tell it to cross around to your side, between the sheep and the fence, and to drive them away again (see Figure 24.4).

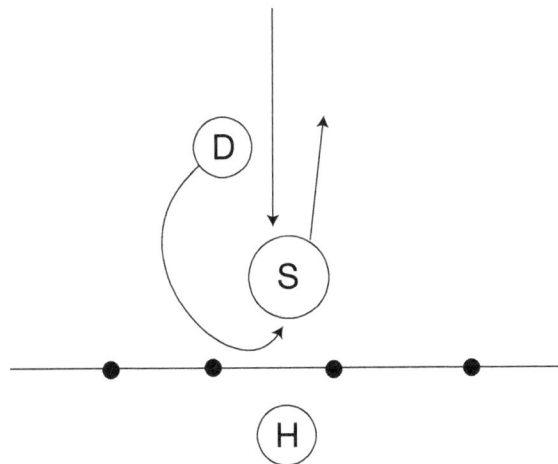

Figure 24.4 Introducing pup to draw ringing

As the pup gets more pliable in this regard you can dispense with the fence, and have it 'draw ring' the sheep in the open before it gets them all the way to you.

Developing the thinking dog

There are a number of ways that you can then encourage *purpose* and *brains* in a pup's paddock work, once it will drive away well.

The first thing is to get it driving sheep up a narrow laneway (or along a road or a fence) *without any commands*. Endeavour to have it drive sheep away along a laneway or a road like this, by using as few commands as possible. Try to use reprimands ('ahhh', 'hey', or its name) rather than commands, to make it figure out what it should be doing without having it spelled out. If it has the right make-up, it will soon learn to drive the sheep along the lane without any commanding at all.

Once this stage is reached, find a situation where you can cast the dog out through a gate into the next paddock, so that it draws some sheep towards you with the fence preventing it from getting them all the way to you (see Figure 24.5).

Then walk away from the fence yourself and and command it to drive the sheep along the fence to the gate, and then through the gate and back to you (see Figure 24.6). Repeat this until it does it on its own.

Then you can command the pup to drive sheep up a fence line, through an open gate, down another fence line, through another open gate, and back to you (or something similar).

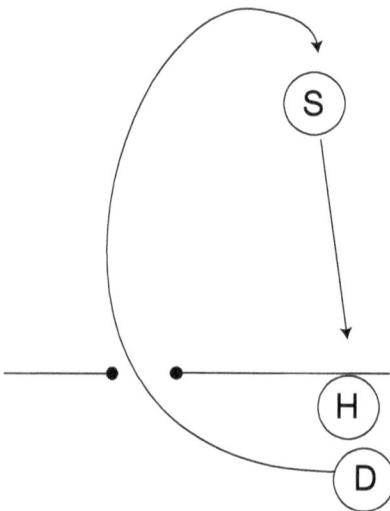

Figure 24.5 Teaching dog to work sheep around obstacles on its own initiative – first step

With a good pup with the right natural ability (especially calmness, drive, and distance) it should begin to do this on its own initiative without commands.

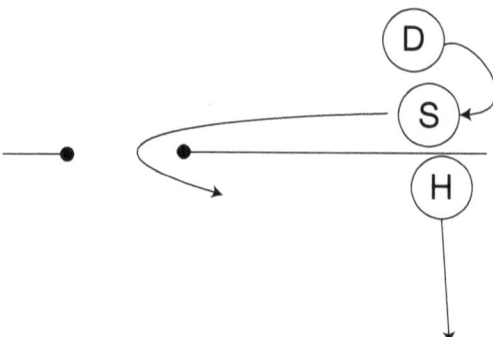

It will learn to look ahead for the gate while driving away, the same as it looks (or should look) for you when it is pulling to you. You can then begin varying the situations, until the good dog will start to look for open gates and the like. It will start to realise that when it is told to drive sheep away that it is trying to do something with them, and will begin to think for itself.

Such dogs (and there are very few) can learn to take sheep around obstacles in their path without being told, but an

Figure 24.6 Teaching dog to work sheep around obstacles on its own initiative – second step over-heading, busy or excitable dog never will, no matter how much training or experience it has. Neither will a dog with too much eye that has no time to think about anything except what it is staring at.

'Round (or Over) here'

The 'here' command is added on to the directional commands, or used on its own, when a dog is holding sheep to you, in order to bring it from a pulling position to a driving one (that is, driving with you, not crossing between and driving away).

A good idea of when such a command might be used is if you have cast the dog out and it has brought a mob of sheep to you, and then you want it to come around to your side and drive the sheep with you ('round here'). You can also use the command when driving sheep along a road, after the dog has been sent 'go forwards' (to get the lead moving) to call it back 'here'.

One way to give the pup the idea is to have it drawing a small mob of sheep along a fence behind you. Then call it 'round here' to squeeze between the sheep and the fence around to you, then drive the sheep away up the fence. Then send it 'go back' and repeat the exercise. Or just teach it during practical work.

25

Recasting and mustering

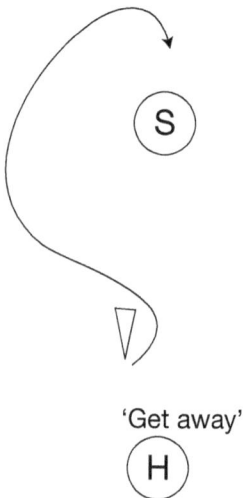

You can start to teach the pup recasting on one mob early in its education, particularly if it is a natural caster. You are simply casting it from between you and the sheep, with it facing *away* from the sheep, instead of the other way around.

You walk away from a quiet group of sheep with the pup, and stop it facing away from them. Then stand outside the pup, so that it is between you and the sheep, facing you. With a wave of the correct arm (right arm if you want the pup to go clockwise around the sheep) and the command 'get away', you will readily get it to turn around and cast around the sheep (see Figure 25.1).

If the pup has good 'square break' and 'width', and naturally breaks out from this position to curve wide around the sheep, it won't take much effort to teach. But if it goes close and straight around the far sheep, and shoulders them, you will have to take much longer to teach it to break out properly, if you want a well-finished dog. In that case, once it turns around make it 'sit down', and then use 'get out' to make it cast around them properly.

Figure 25.1 Introducing 'get away'

Later work recasting – 'get away', 'get out', and 'come in'

When you really get serious about teaching the pup to recast off one lot of sheep to go for others, there are two main ways you can go about it. The first is to teach it to split sheep (see Chapter 28), so that you can easily create two mobs to work with out in the paddock. This is the method I prefer, because it allows you to more easily approximate practical working conditions, and because teaching to split sheep is a valuable exercise in its own right.

The other method is to use two adjacent yards in the sheep yards, with a gate between the two. With 10 or 20 sheep in one yard, you put half of them through the gate with the pup, and leave half of them behind.

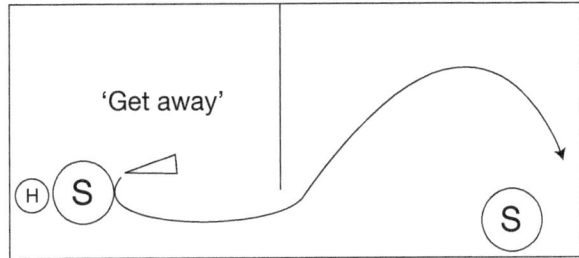

Figure 25.2 'Get away' in two yards

Move around to the far side of one lot of sheep so that the pup is now holding them to you, with its back to the open gate and to the far lot of sheep in the next yard. You should readily be able to send it 'get away' and so put the two lots back together again (see Figure 25.2).

From then on you simply do the same exercise in bigger yards, and finally out in two adjacent paddocks joined by a gateway.

However, if you teach the pup to split sheep you can more easily create varied situations, and position the far lot of sheep 'blind' and so on. Just because a dog will *cast* blind does not mean that it will be so willing to leave sheep to *recast* blind.

Some people worry that teaching a dog to split sheep can become a bad habit. But in fact it is more the opposite, provided that you teach 'get away'. If a dog is a natural mustering dog then it will still be a natural mustering dog, even after you have taught it to split. If it isn't, then it is in even greater need of this training. In fact, if you overdo the work on 'get away' after the pup has split the sheep you can have trouble stopping it from going back for the far sheep.

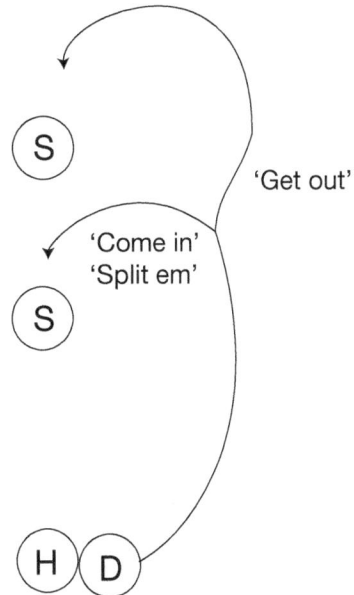

Figure 25.3 'Get out' or 'come in … split 'em'

When training, if you have taught your pup to split sheep, you can position two mobs with one some distance behind the other, and practise with the pup casting to either 'come in' around behind the first lot, or 'get out' to keep going around the second lot (see Figure 25.3).

Or you can position one lot well out to the side of the main mob, and send the pup around the main group; as it casts out make it 'get out' and break out around the other group out to the side (see Figure 25.4).

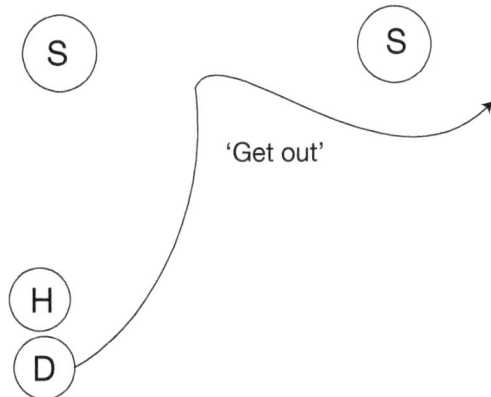

Figure 25.4 'Get out' with one lot out to side

Campaspe Bounce set up for a straight cast

These exercises are really the finishing touches on teaching 'get out', as the pup learns to look for more sheep. If you put the second lot of sheep out of sight some distance away, it will learn.

'Get away' and straight casting

Sometimes it is beneficial to be able to send a dog in a straight line towards where sheep are to be found, and only break it out around them when it gets close.

For example, in rough, hilly country, if you cast the dog out wide it may have to go right around impassable rocky hillsides, right out of its way, before it can get to the sheep. But if you can send it in a straight line, along a sheep track, say, it will save an enormous amount of time and energy, and be more likely to find the sheep. Then when it gets near them it breaks out around them.

You can teach this in two paddocks with a gate in the middle of a fence line between them. Position the pup between you and the gate, and make it 'get away' directly for the gate. Once it is through the gate have it cast around the sheep (see Figure 25.5).

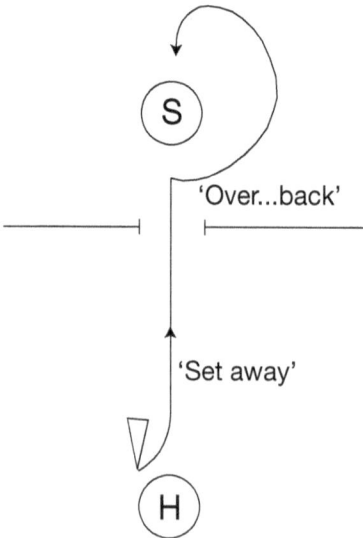

'Over...back'

'Set away'

Figure 25.5 Casting straight, then break out

'Here away'

The next useful mustering command is 'here away'. This is to pull a dog off sheep that it is driving, and come back behind you to retrieve sheep that may have been missed or dropped off behind you, such as when coming through scrub.

Once again you can teach this in two adjacent yards, with the pup holding one lot of sheep against the far fence, and you standing somewhat behind and to one side of it. Call it 'here...away' to come back past you and retrieve the lot in the yard behind you (see Figure 25.6). Or you can teach it on two mobs separated in the paddock.

'Get over' or 'Go round'

Even with these two commands 'get away' and 'here away', there are still other mustering situations with scattered stock where neither is applicable.

For example, when mustering in rocky or scrubby country, sheep often split up around boulders or hills or scrub and so

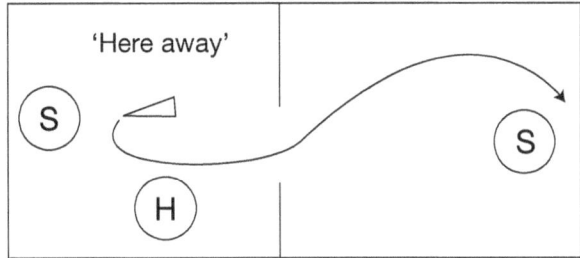

Figure 25.6 'Here away' in two yards

on, and you can end up with pockets of sheep in various places. In these types of situations you can need to tell the dog to recast sideways in relation to you, rather than directly away from you or back behind you, and so 'get away' or 'here away' don't fit the bill.

In such cases, if you give a *directional* command the dog will generally obey in relation to the sheep it is working, rather than breaking off them and going for others it can't see. So I do an exercise at home to teach it to break off sheep that it is driving across, to go for others, in response to either 'get over' or 'get round'.

Work on a fence, and have the dog split some sheep up, and drive one lot away up the fence, while you remain in the one place. Use the crossing commands 'come round' and

Campaspe Bounce working among the rocks, 10 months old. He is driving the sheep across me to join with others further to the right.

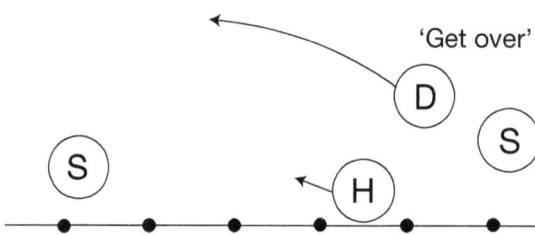

Figure 25.7 'Get over (or round)' to recast sideways

'come over' to drive them away. Then when the dog is in the position as in Figure 25.7, tell it to 'get over' and make it kick out around the other lot of sheep, and vice versa in the other direction. Face in the direction of the sheep to recast to, and walk towards them as you give the command.

Then when out mustering it will know the difference between 'come over' (to cross) and 'get over' (to recast sideways), when driving sheep across you.

Summary

Once the pup understands this work, we should have a well-educated mustering dog that will rapidly develop a sound knowledge of practical mustering as a result. Provided that it is well bred (which is the most important factor), it will become a top mustering dog in no time.

26

Force

When it comes to force, the prime requirement is breeding, as in most things. Ideally you shouldn't have to go to much trouble to teach a dog to force, or to build its confidence.

If a dog is naturally weak there is not much you can do about it, although if the weak dog has a natural desire to force it will gradually gain confidence and improve with general work, but it will never have real strength.

However, if it is one of those dogs that just looks at sheep with no inclination to shift them (like many three-sheep dogs), then it will never amount to anything.

Confidence – step-by-step

It is important not to allow the pup to get knocked about by stock, before it has time to gain confidence. Therefore you should keep it away from ewes with lambs, and particularly rams, cattle, or cows with calves, or bulls. If it is a really strong pup this is less important than if it is weak, but all pups require some time to gain confidence.

The clue is to gradually expose it to more difficult stock as it gains experience, but there are a number of other pointers and methods that can help. Working the young dog with another strong dog can help it gain confidence more quickly, such as when working cattle, but ideally it should be well enough bred not to need any of this support.

Another point is not to ask the pup to force, or indeed allow it to do so, when the stock have nowhere to go – such as up against a fence or in a corner. Otherwise you can teach the exact opposite to what you require, because the pup learns that the sheep go back past or over it when it forces. This will result in your force command becoming a 'show caution and get ready for a break' command, and instead of forcing calmly the pup instead gets tense and begins showing more eye (if it is that way inclined). Therefore, you should only ask the pup to force (particularly a weak pup) when the sheep can move off it in response.

To force a couple of sheep into a corner until the sheep are pushed up against the fence, and the dog's nose is inches from the sheep's, can be a useful exercise later in the dog's training with a strong, steady-forcing and tight-holding dog (I well remember Jack Hiscock giving this sort of demonstration). However, it is certainly not for the young dog or the dog lacking in good, steady force, or cover, or one that is likely to bite.

Once the young dog is reasonably confident, a good exercise is to stand in a small gateway and have it hold and force some sheep through, with you in the way providing some resistance, or do the same exercise over a creek. This gives it the idea of achieving something by its force, and of holding while it forces.

Shed exercise

A very useful exercise to increase a pup's confidence is to work it in a holding pen in the shearing shed, or other small yard, fairly full of sheep. Follow it around the sheep, commanding it to 'go round' or 'get over'.

Because the yard is full of sheep, the sheep around the edge will sooner or later be facing the pup. Therefore it must push past them, because you are behind it preventing it from coming back past you. (With some dogs you may need a rope on them to prevent them from crawling *under* the sheep, or from going *over* them, and so teach them to go *around* the sheep pushing them off the fence.)

It rapidly learns that sheep aren't as frightening as it thought, and that the only way to end the confrontation is to squeeze past them. You can also use a similar situation to teach 'walk in' – in that case, use the rope to prevent it squeezing around them.

This exercise follows the same principles as teaching the timid dog 'here', and will reduce a weak dog's fear of sheep just as it reduces the timid dog's fear of people.

This is also a reasonable test of a dog's strength, to see how it reacts. Rod Cavanagh tells me that when he went to visit Allan Miller, taking a pup with him, about the first thing Miller would suggest would be to put the pup in the shearing shed in a pen full of big, rough wethers, and see how it reacted.

'Walk in'

Firstly a word on the two commands 'walk in' and 'push'. I use 'walk in' for a steady, direct, walk-in force on one point, while with 'push' the dog is allowed (or encouraged) to force and cover more strongly, and perhaps bark.

The well-bred dog should naturally possess good, strong, unhesitating walk-up force. So for some dogs you will not need to give it much work on 'walk in', and it is my opinion that if you do need to go to much trouble then the dog will never be really good.

However, many Australian dogs, being too much on the head and lacking drive, and being excitable and often weak, tend to slide around as they force, and therefore it is difficult to apply force where it is needed.

There are exercises that you can do to improve any dog's force. The first thing is to teach the general meaning of 'walk in'. The pup gets a good start in this when you are teaching it to lift sheep, and to draw them steadily to you, by alternating the 'sit down' and 'walk in' commands. This is all that will be necessary with the strong pup, along with teaching it to 'walk in' when driving.

But it does not follow that the weak dog will 'walk in' when sheep are facing it instead of moving away from it, or when they are stopped. A good exercise for a weak pup is with a few quiet sheep in a corner. Get the pup to 'walk in' towards them as far as it will come, and then make it 'keep off' (walk backwards) for some distance, and then repeat. Try to get it to come in closer each time.

I have said before that getting a pup to force sheep that are on a fence isn't a good idea, but with a weaker pup that won't come in far enough to cause the quiet sheep to break over or around it, it isn't such a problem. But you must have quiet sheep, for that reason.

So ask the pup to 'walk in' once, and only repeat the command if it stops or slows down, provided that you are confident that it will come in again (have a rope on it). But when you see that it isn't willing to come in any further, don't keep repeating 'walk in', because you will teach the opposite, but rather make it move back off the sheep and then repeat the exercise. Each time try to get it to come in a bit further on its own.

Now the main requirement in really teaching a dog to force is in having stock that will stand up to it to some extent, and so give it something to force against, but that move off it gradually as it forces.

One good method is to work the pup on a ewe and a lamb, provided that the pup has the necessary ability and strength to do so. The ewe will tend to stand up to the pup, and so provide something for it to force against, and yet being in the open paddock (or along a fence) there is nothing to stop it from backing off. Another trick is to put some sheep in a yard, and put one sheep over the fence in an adjacent paddock, and have the pup work the single sheep away from the others.

You can eventually do this exercise with one ram, but it is hopeless with a pup lacking drive or short cover or strength, or lacking the ability in general to handle one sheep *while still giving it somewhere to go.*

You can also do the same exercise as described next, except using only one or two sheep instead of a dozen or more, standing in a gateway and having the dog force them through.

'Push'

'Push' is a more vigorous, or more free-moving, type of force, although this isn't to say that it is fizzy or erratic. The good dog works its sheep smoothly, and generally shouldn't bounce or jump or fizz around, whether that is one sheep or a thousand. Most often, the dog that *needs* to bounce or fizz to shift stock is weak, or has too much eye. *Any 'bounce' should mostly come from a dog's high degree of short cover when it is in close to sheep forcing. This type of short cover is very rare.*

The best method is to stand in a narrow gateway or opening, and have the dog force a dozen or so sheep through with you standing in the way, 'push'. Start with your quiet sheep which will come through readily, and work your way up to wilder sheep that will provide more resistance, so that the dog will have to really show some cover and strength and 'block and hold'. You can do the same thing across a creek. Clapping your hands, or making a hissing noise, can encourage *some* dogs to force, but each dog (or bloodline) is different.

This holding and forcing exercise will give you a pretty good idea if your dog is any good or not. If it has a natural cast and lift and draw, and will hold and force wild sheep calmly in this last exercise, and also work and force one fighting ram, then you have a good dog.

This is one of the main downfalls of many UK dogs. Many of them are very strong dogs with good direct, controlled force, but when asked to really force *and hold* a group of sheep which are trying to split and break back over them, they will end up shouldering and grabbing hold instead of forcing and holding and breaking cleanly.

Many Kelpies tend to rush around to force, and not hold while they force, and most lack the cover to hold tightly – they generally move late. Some of them have good break and holding ability, but won't force at the same time – even though they might have plenty of force when they rush, they have none while holding.

Most modern three-sheep show collies simply lack enough strength or force full stop, and aren't even worth considering. However, some of the Yulong Russ blood dogs were (and are) exceptional in their ability to block and hold and force strongly and cleanly.

It is this ability that makes the best forcing dogs – the combination of exceptional cover and holding ability, and strong, direct force applied simultaneously.

Bark

It is easy to teach excitable barking dogs to bark. Simply let a few other dogs off to run around and the dog left on the chain is likely to start barking, whereupon you can put a command on it. Then alternate between 'speak' and 'quiet' until it understands. Then have it 'speak' every time before you let it off the chain, or before you give it a feed, and it will quickly respond.

A weak barking dog will bark when stock face it and put pressure on it, so it is only a matter of setting up such a situation and putting a command with it.

Another method is to have the dog hold some sheep in a corner, such as in the forcing yard behind the draft, or in a race leading up to the shearing shed. Stand between the dog and the sheep and clap your hands and command it to 'push, speak', but block it from coming in. The idea is that it will become frustrated and begin barking. (Some dogs will bark to the crack of a whip, or even if you just clap your hands a bit, but such dogs are often a bit ratty.)

These ideas will not work with a really calm dog, who is likely to quickly decide that you are not in your right mind, and just stand there looking at you as though you're a fool. To get a calm, strong dog to bark it must possess that natural calm, purposeful bark itself.

(This reminds me of a story. One time I had two young dogs with me, and both would heel. I was on foot, and had walked down to the creek to start a petrol pump to fill a water trough. As I started the pump, the noise triggered both dog's forcing instinct and they dived in simultaneously and heeled the nearest thing – me – one on each heel. Needless to say, neither of them was terribly popular for some time.)

If we command the dog to bark at the right times, and make it 'be quiet' at all others, it will learn to use bark at the right moments on its own (provided that it barks calmly and not out of excitement or weakness). However, if it *always* barks as it forces, then by making it 'be quiet' you can effectively stop it forcing completely. (The same applies to biting dogs – if you teach them not to bite when forcing in the yards, or on a mob, then generally you stop them forcing completely; the only solution is to muzzle them.)

Summary

Force is one of the most important aspects of the practical working dog. It should force directly, and not rush up the wings and so forth except in rare cases. This is a big failing of most Australian dogs, and a big plus of the UK-bred dogs. But very few dogs really hold while they force, because they lack a high degree of short (and mob) cover, combined with strength.

Basically, if you want a top dog then it should force naturally without any real training at all, and this force should be balanced by distance and calmness, so that it forces when required and then backs off and pokes along quietly. Force should also be combined with short cover, break, hold together, and width, so that it holds while it forces.

27

Yard work

The requirements for yard work are vastly simpler by comparison to those of the high quality mustering dog, which is one reason why yard trials have gained such popularity. Not only is it much easier to breed a competitive yard trial dog, but the level of handling and training ability required is also fairly minimal.

There are two main areas of additional training needed to turn the paddock dog into a finished yard dog. The first is teaching it to work the race, and the second is teaching it to back sheep.

A holding yard with 1000 merino wethers waiting shearing

Working the race

If you have taught the dog 'go forwards' in the paddock, you can quickly teach its meaning in the yard – to run up the side of the handling race, using 'go forwards' and 'push' or 'speak'. Have the dog come back towards you with 'to me', and 'steady' or 'stop...push'. It is better if the dog works the race as it comes back towards you, and doesn't simply run back to you.

If you haven't taught 'go forwards' in the paddock, simply use it in this situation until the dog understands.

Backing

Some dogs like backing, and others don't. I have seen Kelpie pups no more than three or four months old, which you had a struggle to keep on the ground in the open paddock with quiet sheep; all they wanted to do was clamber on top of the sheep. So some dogs require very little training to back at all (in fact some need training to stay on the ground), while others might never be much good. The good backing dog has good balance, and tries to remain on top of the sheep, whereas the dog with less aptitude is always looking for an excuse to get back on the ground.

The main requirement when teaching to back is tightly packed sheep, with as much wool on them as possible, to make the most stable platform for the dog to move on. With such sheep, in a small catching pen or a race, you shouldn't have much trouble in teaching the dog the meaning of 'hop up', with a lead on it. It should already know this from jumping up onto the ute, or bike, or over a fence.

Once it is doing this readily, you can begin to stand on the far side of the sheep from the dog, and call it to you with 'hop up, here', and then send it back with 'go back'.

You should teach it the difference between 'go back' and 'go forwards' easily enough. With sheep packed in a race or in a corner of a pen, send it up onto their backs with 'hop up...go back', and make it jump over the far fence on the far side of the sheep, down onto the ground (although later it should learn to hop down without jumping over the fence – have a rope on it). Once it understands this you can change to 'go forwards', to send it on the backs to the far side of the sheep without getting down in front of them.

Then you can introduce it to 'speak' on their backs, and also to be able to move it around with 'round' and 'over' so that you can position it where required.

Remember, however, that backing is another thing that is often overdone. In many cases, such as penning up, the dog is generally far more effective on the ground getting the lead moving, instead of jumping up on the backs and going to the far side when it is the leaders on this side that need shifting.

Also remember that backing and falling down between sheep and getting trodden on and trampled is hard on the dog, and is best avoided where possible.

'Out of it'

'Out of it' is a very useful command in the yards. I teach the dogs to 'out of it' during yard work, such as when drenching sheep. The dogs are in the back of the forcing pen forcing the sheep into the drenching race, then once the race is full I make them 'out of it' to turn around and go to the back of the forcing pen.

Any time they attempt to come in and force the sheep without being told they get growled at to 'out of it'. They quickly learn, and as soon as I close the race gate they don't need to be told, but just quietly turn and go to the back of the forcing yard and lie down to await the command to force again.

How often do you see people working in the yards, with dogs being allowed to force sheep against a closed gate while their handler is drenching or something like that? It is no wonder some people have trouble moving sheep in the yards – by dogs constantly forcing when the sheep have nowhere to go, the sheep learn that when dogs are forcing there is only one way to go, and that is back past the dogs. And the dogs generally lack the 'block and hold' to prevent it.

If dogs only force when the sheep have somewhere to go, the sheep learn to look for somewhere to go, and become good to handle.

As soon as the sheep are flowing make your dogs back right off. Otherwise the sheep simply become immune to force for one thing, just as they become immune to constant barking, and they often get jammed by being forced in too tightly and so won't flow. Use the minimum force necessary to get any job done, and remember that:

An ounce of force in the right place is worth a pound of force in the wrong place.

In fact, a pound of force in the wrong place (or at the wrong time) is often worse than if the dog hadn't forced at all.

28

Splitting and working one

Teaching a dog to split sheep is fairly uncommon in this country, although it was more common years ago. However, I consider it a very useful skill. For starters it is the best way to set up situations in order to teach the pup the various mustering skills. But aside from that it also has many practical uses. Some of them are outlined below.

- To split a few from a mob, in order to put them over a creek or into a difficult set of yards to provide a decoy to help put the rest of the mob over or in.
- To help the dog understand its role when you want it to hold a single sheep off the others, such as when catching a ewe in lambing difficulties, or cutting a bull out.
- To give it the idea of cutting in and blocking one mob in order to prevent it becoming boxed with another mob. For example, in rough country you can come over a ridge and find other stock in your path, and quick action can be required to prevent trouble.
- It helps the dog to understand to work only the cattle, or only the sheep, in a paddock with both. This is not such an issue with a dog lacking 'hold together' or mustering ability, because such a dog readily cuts in among the other stock, but a good dog will tend to keep out around all of them.
- When drafting stock in the paddock, such as separating a few straying cattle from another mob. If the dog has some idea of splitting it will understand to help you cut the ones you want off, instead of working against you and trying to hold them all together.

Teaching to split sheep

Teaching a pup to split sheep is fairly simple. Before you teach this though, it needs to understand driving away, and be well trained and obedient generally. This is probably at about 10 months old with a good pup.

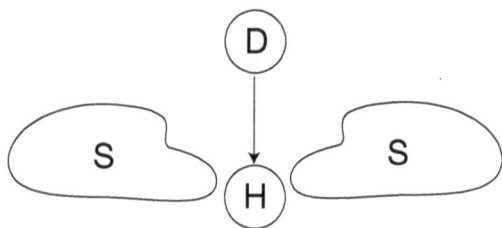

Figure 28.1 'Here' then later, 'split 'em'

I start with the pup holding a small mob of quiet sheep, perhaps 20, to me on a fence in a paddock. I position myself halfway along the mob, standing between the mob and the fence with my back to the fence, and the pup opposite. I then stop the pup and walk towards it to split the sheep myself, and call it in to me with 'here', before having it 'walk in' to drive one lot of sheep away up the fence (see Figure 28.1).

I then either send the pup to head them with 'go back', or bring it back around the other lot with 'here away' or 'get over (or round)', and so in one way or another put the mob back together, and then repeat the exercise. It is good to vary these two ('go back' or 'here away') so that the pup doesn't start to anticipate and learn to head sheep that it is driving when it shouldn't do so – keep it guessing. I also always make it 'sit down' before either of these commands, so that it learns never to head or leave sheep unless told to.

Gradually the person does less of the work splitting the sheep, and the pup comes through on its own in response to being stopped in the required position and called through, initially with 'here', and then once it has the basic idea with 'split 'em'. It will soon learn. Weak dogs with too much 'pressure eye' will pose the greatest problems.

Teaching to cut one sheep out

Teaching to cut one sheep off the mob is simply an extension of this work, but using a separate command so that the pup knows what we want. Begin in the same position as when first teaching it to split sheep, but gradually position yourself and the pup so that it cuts fewer and fewer sheep from the end of the mob, until eventually only one is cut off.

A different command is gradually phased in for this work, such as 'this one'. It later becomes a useful command to cut a fly-struck sheep off, or a lambing ewe in trouble, or a bull out from the cows.

Before this stage is reached though, you must have given the pup some experience in working one sheep, and not many have the necessary ability or strength to do so. So only the better dogs will be able to achieve this cutting and holding one sheep off.

The 1920 book *Management and diseases of sheep in Australia* said, with regard to training the young Kelpie:

> When he has become thoroughly accustomed to sheep, and has reached his full size, which should be at 9 to 10 months old, it will be necessary to teach him to become self-reliant, and to work sheep by himself both in small mobs and singly.
>
> As these lessons go on the number of sheep may be reduced until a single sheep is successfully worked. The dog may then be trusted to work in the open paddock. (Cramsie *et al.* 1920)

Yet how many dogs in this country nowadays can work a single sheep well? Very few, but there are some. (The UK dogs are generally much better at this.)

I remember one day working Cummins Creek Chocker, and droving a mob along a fence line. There were sheep in the paddock I was going through, and sheep over the fence also. For some reason, one wether jumped the fence and took off for the sheep in the next paddock. Chocker leapt the fence and outpaced it and cast around in front of it, and blocked and held it from a distance, then gradually turned it and worked it back the 100 metres or so towards me. I hopped over the fence and Chocker worked it right up to me so that I only had to reach out and take hold of it, and lift it back over the fence.

29

Working more than one dog

When working more than one dog, the main requirements are temperament, natural ability, and control. With good, calm, steady dogs with the right natural ability that require very little commanding, it is not difficult. But with pushy, excitable, busy, over-heading dogs with too much eye that require constant commanding to keep right, it can be a real handful.

The UK system, of working two dogs with two separate sets of commands, is not much use for practical work with a comprehensive commanding system, because there are too many commands to duplicate. And it isn't much use if you have more than two dogs you might work together, because then you might need half a dozen sets of commands. So I use a simpler system, not so suited to show work, but I think more practical.

The first thing is to get control, and the basis for this is 'sit down'. I first teach the dog to remain lying down while other dogs are running around in the backyard. Then I get two dogs and do a little exercise with them, sitting them both down, then calling one to me, focusing all my attention on it and using its name, 'here'.

The other dog must be made to realise that it must stay lying down, unless I focus all of my attention on it and call it directly. Any other command, unless I am looking directly at that dog and using its name, is to be ignored.

I extend this until I am putting one dog through its paces in the backyard, 'stop there', 'here', 'heel', and so on, while the other must remain lying down.

Then I repeat this exercise on sheep, working one dog and keeping the other one lying down. (It is important to have the pup trained to remain lying in place for extended periods of time at home first, otherwise it will get fidgety quickly.) One useful trick when teaching a pup to remain lying down for extended periods is make it lie on a coat or jumper. Anytime the pup moves off, make it go back to the coat. This gives it a point of reference and helps it to understand what is required.

The next stage is to work first one dog and then the other on the sheep, alternating regularly between the two, making the other one 'sit down', and remembering to use the dog's names before allowing them to move off from 'sit down'. If I am working two or more dogs and I tell one to 'sit down', I do not allow it to move off unless its name is called.

Other than that I simply use each dog's name before any command – with experience they will get the hang of it.

Heel

Teaching your dogs to remain at heel while you work another dog is also useful for practical work. You may want to keep one dog at heel while you cast your best dog out around a couple of stragglers on the hillside, or keep one dog at heel while stock are fresh until they settle down, or to force a dog to take a breather on a hot day.

I suggest starting on this once the dog understands the above work, and you shouldn't have too much trouble if it is going well on that. Experience from then on is the key. But unless you have good calm dogs you will never achieve a high standard at this sort of multiple dog work.

The following gives some idea of what should be aimed for:

> I opened the gate in the corner of the paddock and commenced to drift a few cows through. This was easy at first, but as the number of cows left with the calves dwindled, the job of guarding the gate, cutting out and allowing only cows through, became increasingly difficult. Working at it quietly and patiently for an hour or more, I eventually shut the gate on the last cow.
>
> Thinking about it I decided that perhaps few men had ever drafted 140 station cows and calves single-handed. But then the whole fell into its correct perspective as a classic example of what we have been discussing all along: correct dog training (my dogs had all remained patiently waiting up on the hillside except one heading dog quietly holding the mob) and working stock quietly, carefully and firmly with self-control and in this instance a lot of patience. (Mills, McIntyre and Herbert 1964)

In order to achieve this type of dog control, temperament is vital. You can't do this with excitable, ratty dogs.

30

Overview of training

Rearing

a   Tie up
b   Teach 'here' on long rope
c   Develop keenness on sheep, and evaluate natural ability

Early sheep work

a   Break in to 'go back' in yard
b   Use correct arm signals right from an early stage
c   Introduce 'behind'
d   Teach 'sit down' dry (without sheep), then wet (with sheep)
e   Use 'sit down' and 'here' to finish lesson

The directional and positioning commands

a   Introduce 'round' and 'over' in yard
b   Move these lessons out into small paddock
c   Begin short casting
d   Get more control of 'behind', 'round' and 'over'
    i   Reverse counter-balance exercise
    ii   Stationary directions
    iii Over-taking directions
    iv Separate arm signals and vocal commands
e   Introduce the steady lift 'walk in' and 'steady'

Cast, lift, and draw

a   Teach whistle for 'sit down'
b   Teach 'get out' and 'come in'
c   Get control at top of cast and on draw at long distances

Heel

a   Teach pup to heel, then heel it to and from lessons, and when walking back to cast it, etc.
b   Later on, get control of 'heel' while working second dog

Driving

a   Teach to drive with handler (crossing behind), using the long cord 'walk in'
b   Refine use of directional commands, so that the dog doesn't change from a driving role unless 'back' is given, and doesn't change from a pulling role unless 'here' or 'come' are given
c   Teach the crossing commands 'come round' and 'come over' in yard
d   Get pup driving sheep out in paddock until it can drive sheep anywhere in the open
e   Teach 'round here' and 'over here' to come from pulling to handler

Recasting and mustering

a   Introduce 'get away' early in dog's training on one mob, to cast from between you and the sheep
b   Teach 'get away' on two mobs, either by splitting a mob or in two adjacent yards or paddocks
c   Teach 'get out' and 'come in' on two split mobs
d   Teach 'here away' on two mobs
e   Teach 'get over' or 'get round' to recast sideways on two mobs, when driving across

Inhibiting commands

a   Get control of 'stop there' stopping dog on feet if desired
b   Teach 'sit down' for extended periods if haven't already done so
c   Teach lesser 'stop' on sheep
d   Teach 'steady' on sheep
e   Teach 'keep off' to walk backwards if required
f   Teach 'out of it' to turn and go directly away from sheep

Practical paddock work

a   Teach dog meaning of 'go forwards'
b   Use 'get over' or 'go round' for big movements, and 'over' or 'round' for smaller ones

Force

a   Teach dog to walk in
b   Teach to bark on command
c   Work dog around full pen of sheep to increase confidence
d   Teach dog to force sheep through small gate when you are standing in the way

Yard work

a   Teach to work race, using 'go forwards'
b   Teach to back, using 'hop up'
c   Teach 'out of it'

Splitting and working one sheep

a   Teach dog to 'split 'em' on fence line, then later in open
b   Give dog practice working one sheep
c   Then teach it to gradually split fewer off until splitting only one off the end – 'this one'

Working more than one dog

a   Do exercise dry with two dogs, one remaining in place while you mess about with the other one
b   Graduate to making one pup 'sit down' in one place while other dog works sheep
c   Finally work both on sheep, alternatively stopping one and working the other
d   Have one dog remain lying down during entire time you are training another dog on sheep
e   Graduate to practical work

Part 5
AT WORK

31

Handling stock

When handling stock, the first thing to realise is that the same principles apply as when training or handling a dog. You (and the dog) are teaching the stock to respond in a certain way, and they respond rapidly.

Sheep respond more quickly than cattle, because they are generally more timid and sensitive to what the dog is doing. Cattle are often more curious and less timid, and so respond more slowly. It is like the fact that a soft, sensitive dog can be taught various things by certain methods more quickly than a hard, strong dog. This does not make one more intelligent than the other, just more easily influenced.

The speed at which sheep respond to the dog is evident at any three-sheep dog trial. A good dog and/or handler can quieten stock rapidly, and have them behaving well in a couple of minutes or even nearly right from the outset. A poor dog will have 'bad sheep' for the whole trial, *just as a poor handler will have badly behaved dogs*. So when working stock, the dog and handler are correcting certain actions on the stock's part, or rewarding others, and the stock will behave accordingly.

The natural ability of the dog is paramount – where it positions itself in relation to the stock, and what it does there, determines the effect it has on them.

In most of the following discussion I refer to sheep. However, the same principles apply equally to cattle. The only difference is that weakness in the dog becomes an even greater liability.

Working a single sheep

Perhaps the best illustration of these principles is when working one sheep. The principles of working one sheep are exactly the same as working a bigger mob, only the effects are more exaggerated and the failings of the dog become much more obvious.

When working one sheep most Australian dogs nearly always end up too far to the head of the sheep. What happens is that whenever the sheep attempts to move, the dog blocks it, and so punishes that action. The sheep turns another way and again is punished by the dog through it being too far to the head, that is, in front of the sheep.

Pretty quickly the sheep gives up, because nothing it has tried has had any beneficial results. Therefore it must try some other tactic and it either lies down and plays dead and refuses to budge, or stands and refuses to move, or begins to fight the dog, or runs straight over it. Then it often intimidates the weak dog and manages to get away. One way or another you end up carrying it.

Then you have dogs with a bit more drive that attempt to drive the sheep where they want it, from behind, which is good. The problem is that they generally lack enough short cover and balance, so are 'loose' in their work.

Therefore, the dog moves late, after the sheep has already moved a metre or two, and so the dog then has to get further to the head in order to straighten the sheep back into line. It also often overruns. So it is correcting the sheep harder than it should. The sheep then generally does the same thing in the opposite direction, that is, it moves the other way, the dog moves late and has to run around and straighten it up on the other side.

So although the dog is trying to drive the sheep in the right direction, it lacks the balance or short cover necessary to do it properly. It moves late, and then generally overruns. You end up with the same effect as before – the sheep is effectively corrected for every movement, and will soon baulk or fight. It will take longer before it gets to this stage, but it will get there.

The dog with good short cover and balance

On the other hand, the good short covering and balancing dog, with good 'drive', will get behind the sheep and drive it forwards, and is so responsive to the sheep's movements that the sheep might only move five centimetres (two inches) one way, and the dog has already moved over six centimetres (two-and-a-half inches) to cover it and keep it on line.

Because the sheep has an option (to move away from the dog without the dog then appearing in front of it) it will travel along nicely. *It doesn't feel cornered.* It may try to break away every now and then, but the good dog will be waiting for that.

This is why stop-start stock work is mostly undesirable, whether that be a mob (as described later) or a single animal. The poorer dog corrects the sheep late, and must therefore use more pressure and get further in front of it and so corrects it harder, and mostly gets much worse results. The good covering and balancing dog instantly corrects the sheep's slightest deviation in the most tactful manner, *while leaving the desired direction open*, and achieves its goal.

Also, a dog that rushes around to force, instead of forcing directly on the point of balance, creates problems whether working one sheep or a mob.

A good handler training a dog is exactly the same as a good covering dog. They read the dog's thoughts, and react to them by instantly and finely correcting the dog almost before (but never *actually* before) it has done something. The poorer handler reacts late and corrects late, and must therefore correct more harshly to get the dog back on track.

The manner in which I have described a poor over-heading dog working one sheep is the way in which some of the more successful three-sheep trial handlers deliberately handle their sheep. They breed or obtain dogs lacking in balance or eye or drive, and they then move the dog around by command, first to the tail to allow the sheep to move, then to the head to block them, and then back to the tail, and then up to the head, and so on.

In this way they correct the sheep every time they try to run, and so quickly slow them down. Three sheep don't feel as threatened or as cornered as one, because they have the moral support of their mates, so they don't lie down or baulk or panic with this treatment as easily as one sheep does. As the sheep quieten down the handler keeps the dog back off the head, and so gives them somewhere to go.

The other effect of this type of tactic is that because the dog appears to the sheep to be here, there, and everywhere all at once, they give up trying to get away. The handlers choose dogs that are very fast (and excitable and busy in order to give fast response) and lacking in eye and balance, and then work them in this manner. But it is the handler working the sheep, not the dog, and such dogs are useless when left to their own devices (as you must in real work), or on wild sheep.

The best dog, the natural dog with great balance, works by itself without any of this mechanical 'back and forwards' rubbish, and far better than any handler ever could, and will do it even out of sight over the next hill.

Working a mob

Exactly the same thing happens when a dog is working a mob as when it is working one sheep, only the effect is less exaggerated. If a dog is constantly getting too far up the wings towards the head, the mob may not all lie down and baulk, but they will become doughy and difficult to move. They won't flow.

A good example of how a dog getting all the way in front of the mob will slow it down is if you let a mob of sheep go back to their paddock down the road. Sometimes the leaders will travel along very quickly and perhaps leave the tail behind. If you send your dog to block the lead and turn it back into the others, then call your dog back to the tail, and then do the same again when the leaders take off, generally that is all that is required to convince the sheep to go a bit more steadily. The dog has corrected them, and they will go more slowly. (This is like some trial handler's mechanical head/tail technique.)

However, you sometimes see stockmen working a mob as if they were working the side of a drenching race. They send the dog three-quarters of the way up to the lead, and then call it back, in order to get the mob moving more quickly. The problem is that they do it constantly, over and over as their stock-in-trade for moving sheep. This may work in the short term, and is an indispensable trick at times, *but in the longer term it only makes the stock doughy*. It is the same effect as when sending the dog all the way in front of the mob to block it, only less obvious because the dog doesn't get in front of all the sheep. (Out in the paddock, however, it is sometimes necessary to have dogs working out on the wings to shepherd (balance) the mob in the desired direction, but this is different to sending a dog up a wing close to the sheep and then calling it back in order to speed the mob up.)

Advantages of working sheep from behind

Another good example of why the good dog must naturally work sheep from behind, is when you cast a dog out around a small mob, say 100 or so. Most dogs when drawing the sheep back to the handler come too far around the wing, and so turn the mob, and then have to go back around the other side to straighten it. But in so doing they come too far up the wing again and turn it back the other way. So the wing keeps ringing around the lead and slowing or stopping the mob, and the tail breaks back behind the dog. Instead, if the dog drove the mob from behind this wouldn't occur, and the lead could lead along nicely. Thus you can see that working a mob follows the same principles as working one sheep.

The fastest way to move stock in the long run, if they have any real distance to go, is to allow them to go at their own pace as much as possible, and to work them from behind.

'Let the lead lead, and the tail draw along.' It is a case of: 'make haste slowly.'

If the sheep stop, then by all means get in and shift them, but as soon as they respond, back off. Correct them for stopping, reward them for going. If your dogs keep pushing into them after they have moved off, then you are effectively punishing them for stopping, and punishing them for going. They will feel cornered because they feel that they can't avoid the dog, and so they will become doughy (this is an example of 'learned helplessness').

This reinforces the importance of steady dogs with distance (which comes largely from a calm temperament and drive). The good dog forces when required, then backs right off and eases the pressure when stock respond, exactly as it should.

Disadvantages of working with motorbikes

This principle is one reason why you can't work stock well off a motorbike. The noise of the bike is effectively punishing the stock, because they don't like it and are trying to get away from it. So, in order to ease the pressure on the stock if you are riding a bike, you have to get a long way back off the stock, or switch the engine off. You then lose contact with the stock, and so the bike is less than ideal, particularly in rough country where you must stay in close contact.

You end up with stock running hard, and often weak sheep lying down, and then the stock get knocked up and won't travel. You are punishing the stock when they stop, by honking the horn or revving the engine, but then still punishing them when they move off because the engine is still running.

If you are on a horse the stock generally more or less ignore your presence, and will travel much more quietly and calmly, so in the long term they will also travel more freely. If you keep them calm they will have more stamina, just as a calm dog has more stamina than an excitable one.

Also the noise of the bike excites most dogs, and causes them to work faster than they would with a horse, and so the dogs knock up more quickly. In fact the engine noise becomes a force command, because the dog learns to force when you rev the engine and honk the horn, and then when you simply speed up on the bike the dog takes that as a signal to force. This all results in a triple effect of not only the noise of the bike chasing the sheep, but also dogs working at twice their normal pace, *and* forcing excessively.

Droving or pulling?

Something sometimes suggested by dog breeders, when confronted by farmers complaining about dogs always wanting to get in front of the mob, is: 'Why don't you go in front and let the dogs bring the sheep behind you?' Then the breeder adds the standard line about it being natural for a dog to want to head sheep, and to go to the opposite side of sheep and work them to the handler.

The breeders are right, and they are wrong. It may well be natural for *their* badly bred dogs to always want to head sheep, and to have no idea of driving them. But most likely their dogs have no real 'natural pull', despite what they say about heading dogs naturally wanting to bring sheep to the handler (precious few actually do) – in general the dogs go to the head simply because they want to head the sheep, due to an overdeveloped heading instinct made worse by excitability and busyness, and by a lack of drive.

Some breeders and trainers also advocate the handler leading a mob simply as an excuse for poorly trained dogs. The first thing they teach the dog is to go around the far side of the mob (this is generally the very first thing most handlers teach when breaking a pup in). You make it unpleasant for the pup to be around your side, by growling at it and chasing it around the far side, and then leave it alone when it is over there. Then people wonder why it always wants to be on the far side, and say it is natural for it to do so. *After this is taught, you then have to teach it to drive with you.*

They are also wrong about the handler going in front of the mob and letting the dog bring the mob behind them, because most probably the dog in question lacks the ability to be trusted with the task. It probably can't be relied upon not to leave sheep behind, or not to rush the stock, particularly if there are a few weak stragglers or the country is rough.

Jack Hiscock with a mob of about 500. (Photo courtesy Rod Cavanagh.)

If it won't drive stock WITH you well, then neither will it drive stock TO you well, because pulling stock is simply driving them to you. The only difference is your position, not the dog's actions.

The handler's place is on the tail or flank of the mob, or a wing in some instances, where it has always been, where he can see what is happening and work the stock to their best advantage. Letting a dog hold stock to you may be the best way to work small numbers of touchy sheep or cattle with a good dog, but certainly not with a mob of any size in most situations.

Ewes and lambs

Handling ewes and lambs is where a stockman's (and his dog's) inadequacies often become apparent.

The secret here is to let the stock go at their own pace, and to keep the ewes and lambs mothered up. You often see poor dogs and handlers rushing around and bouncing the mob (sooling them), and then the ewes and lambs get separated. The stock then have to be forced every inch of the way home because the ewes are always turning back the way they have come trying to find their lambs. If, on the other hand, they are allowed to go at their own pace, and to remain mothered up, they will get home in half the time and without the stress. 'Make haste slowly'.

But you can't do this in rough country without good dogs. It may seem to the first type of 'stockman' that the job is getting done more quickly, with dogs rushing around and barking and sooling, but such is not the case.

When it comes to yarding ewes and lambs, and more force becomes necessary (sometimes a *lot* more force), then it is vital to have dogs with good cover and 'block and hold' and break, which cover while they force. Otherwise you can end up with lambs spread literally for miles in rough country, in one's and two's, and spend a lot of time getting them all back together again (although it is a good test of a dog working these little groups of lambs back in).

Mustering

When mustering stock in small paddocks, or in open country, there won't be much trouble. A handy dog should be able to muster such paddocks on its own, but most cannot. You simply muster the paddock to the gate. However, as the paddocks get bigger and the country gets rougher and hillier and more broken, perhaps with more timber or rock or scrub, then things get more interesting.

Depending on the situation, there are various ways of mustering the country, and you can use the prevailing wind direction, the location of watering points, the position of stock camps and so on, to make things easier and to help ensure a clean muster.

In really big paddocks, or very rough ones of thick scrub or timber, first you generally muster any stock you see, and then take them around with you, and recast dogs out to muster further stock back to your main mob as you go.

Mustering in hill country

In smaller paddocks (but still of reasonable size), the difference is that you generally don't work a mob around with you and muster other stock into that, but simply muster all the stock towards one point (and not necessarily towards the gate – in some instances you might muster directly away from the gate). Once again you use the wind direction and watering points, and also the topography in hilly country, to your advantage.

Mustering into the wind

When mustering sheep, often a good plan is to begin by working along the downwind boundary of the paddock. If you ride along in sight of the boundary fence, but as far inside it as possible, you can start the sheep you find towards the other end of the paddock, into the wind. If you see sheep between you and the boundary fence you cast a dog around them, to muster them and push them across inside you, and thus start *them* into the wind.

The reason for doing this is that once you start the sheep in that direction they are likely to conintue that way, because sheep prefer to travel into the wind. So you start some sheep on their way, then leave them and continue on and start some more, and so on. Once you leave the sheep behind they will generally keep travelling into the wind.

But if you try mustering them *with* the wind, then once you have started them off and then moved on yourself they are likely to stop or even turn back. It is like droving a mob with the wind – if you have a mob of 1000 woolly wethers to take eight or 10 kilometres on a hot day with a strong tail wind, you are going to have your hands full and will have to use every trick in the book. But if you have a nice headwind things will be a bit easier.

A sheep first notices the stockman and dog during mustering

So, once you have travelled right across the paddock you turn and come back again, further inside your original track, until you have mustered the paddock to the upwind side. This is where the dog that can be trusted to work on its own with real 'mustering sense' is of great value. It can be poking along behind sheep half a kilometre or more away, out of sight over the next ridge or something, while you poke a few along yourself. If it sees more sheep, or gets the scent on the wind, the really good dog will naturally muster those on its own initiative.

You can also sometimes use the wind direction when deciding which way to cast a dog out mustering, by sending it out so that it will be upwind of most of the sheep. If it misses sheep that are further out there is a chance that it will get their scent on the wind.

Similar to mustering into the wind is mustering stock towards the watering point or a stock camp, if that is at one end of the paddock. Once the stock get onto the sheep tracks they will more readily continue travelling along.

It is a good sight, when working with someone else out in rough country, to see their dogs poking along behind sheep, taking them in the required direction, and yet their handler is nowhere to be seen.

Putting a mob through a gate

Putting a mob through a gate, with handy dogs and a bit of knowledge, is generally a simple affair and the good stockman makes it look easy. However, it is very common to see people having a great deal of trouble – roaring about on bikes, honking horns, and the dogs (usually far too many) running about all over the place like headless chooks, and sheep going everywhere except through the gate.

When sheep are treated like this they panic, and don't even see the gateway. If they are taken up to it quietly, with no fuss, they will generally walk right through. If worked by good, steady holding dogs that give them no option, and so eliminate any thought of breaking away, the sheep will resign themselves to going where they are directed. But if they are worked by dogs that allow (or cause) them to split up and break, they will be encouraged to try all the more.

Also, over-heading dogs keep ringing the lead off, whereas good, steady, driving/balancing dogs allow stock to lead and flow.

A common mistake is to begin forcing just as the sheep come up to the gate, to try to rush them through. What often happens is that the sheep have been looking at the gate and considering going though, when all of a sudden dogs start barking, horns start blowing, whips start cracking. The sheep associate the gateway with all of this commotion. They have been very effectively punished for getting near the gateway, and for thinking about going through.

What also often happens when the sheep finally make for the gateway, is that a badly bred and poorly trained dog runs around in front of them (or even just too far up the wing) and turns them back, or rings the wing around the lead. Once again they are punished for trying to go through the gateway. *They will think twice before trying again.* The stockman or farmer generally blames the 'stupid sheep', but what he needs is a mirror. He has *taught* them not to go through the gateway.

One of the main reasons badly handled sheep are difficult to put through gateways is because they have been worked by pushy or rough dogs that over-head, and that are badly trained. The sheep are then reluctant to leave the safety of the mob, and so no sheep are game to lead through the gateway.

The same thing often happens in the yards. The sheep are going nicely up towards the shearing shed or something, when all the dogs start barking and so on, and what happens? The lead sheep turn back looking for safety in the mob. But if the sheep are relaxed and unafraid, and allowed to flow, they will mostly lead through without the slightest trouble in ordinary cases.

Force *is* needed at times, and barking is handy when used intelligently at the right times. But mostly these two things are greatly overused, and are mostly used in the wrong place and at the wrong time.

So the first secret is to keep the stock quiet, and handle them 'quietly but firmly', and give them no other option, and 'let the lead lead and the tail draw along'. Good balancing and covering dogs, strong, steady and sensible, work stock to a high standard. They may not look to be doing much, but they are doing everything right.

Good stock-work is devoid of FUSS.

Once the sheep begin to flow through the gate make the dogs 'sit down' or back off 'out of it', and let the sheep flow through quietly. The dogs will soon get the idea and will then do this themselves as soon as they see the sheep begin to flow through a gateway.

Difficult gateways

When you are confronted with truly difficult gateways, or putting stock into difficult yards, there are a number of tricks you can employ. Most of them depend, however, on having good and well-trained dogs.

One way is to work the mob up along the fence line that contains the gate, and then, as the lead nears the gateway, send a dog up to the shoulder in order to ring the sheep there around in front of the lead, and so through the gateway. Of course it is important that you can stop the dog before it gets too far to the head and turns the whole mob back, or that it has enough experience (and breeding) to stop itself at the right point, and timing is essential.

A good plan is to send the dog only so far up towards the lead, and stop it there for a second, and the sheep in front of it will speed up forwards, then send it on again and use the forward momentum of these outside sheep to ring around the lead.

A really good covering and balancing and driving dog, with experience, will perform this job to perfection, moving back and forth as the sheep go through the gateway to keep them flowing. A dog with less short cover or drive will never really excel at this job (or any other, for that matter).

With bigger mobs you can sometimes approach a gateway directly, and as the sheep near it send a dog up each wing to the shoulders, and call them back, and so send the outside sheep up to the lead, where they will meet and hopefully straighten out and go through the gate. This is mostly useful with very big mobs and large gateways.

You can also cut some sheep off and put them through first as a decoy, but this is usually a last resort. The trick usually is to get the mob ringing so that it rings through the gate back onto itself – the mob becomes its own decoy (see Figure 31.1).

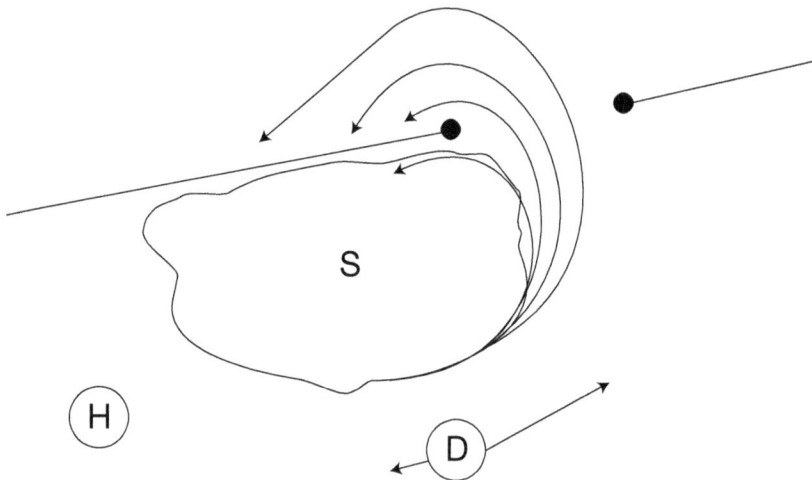

Figure 31.1 Ringing sheep through a gateway back onto themselves

Counting through a gate

To count sheep through a gate you simply stand near the gateway yourself, to count and limit the flow, and have the dogs hold the sheep up to you. Once the sheep begin to flow send the dogs 'out of it' (depending on how big the mob is). It is best to teach the dogs this before you need it, when you can concentrate on them – don't let them come back in until told. A well-handled and experienced dog will get sheep started, and as soon as it sees them begin to flow through the gate it will ease the pressure right off.

With smaller mobs it is probably easiest to count in two's – 'two, four, six' and so on, and keep track of the hundreds on your fingers. Then you don't have any calculating to do at the finish.

With bigger mobs I prefer counting in three's, but some people also count in five's. You count one for each three that go through, so 'one, two, three'. When you count 33 (so 99 sheep have gone through) you count one more sheep to give 100, and keep track of the hundreds on your fingers, or transfer pebbles from pocket to pocket, or call the hundreds out to someone else to keep track of, 'hundred'. Then you start again at one.

Drafting through a gate

To save you having to take stock all the way back to the yards, drafting a mob through a gate out in the paddock can save a lot of time. This can be, for example, when taking the rams out from the ewes – it is a simple matter with good dogs to draft them through a handy gate.

You put the whole mob through first, then stand in the gateway yourself and have the dogs hold the mob up to you. Gradually let the ewes back through the gate, and prevent the rams from coming through, until you have only the rams left.

If you have trouble getting the last few ewes out, you can simply have the dogs hold the sheep up to you and catch the ewes by hand, and manually put them through the gate.

Bringing a few rams home

Putting sheep over creeks

When putting cattle over a creek or river, usually you will not have much trouble – except perhaps with cows and calves and deep, fast flowing water.

With sheep it is a different matter. With a smallish mob and enough good holding/forcing dogs you can sometimes simply crowd them right up to the water's edge until the furthest sheep are physically pushed into the water. However, most dogs lack sufficient cover and 'block and hold' and strength to hold a mob this well – the sheep tend to keep breaking back past them and so they lose any ground they had gained.

Dorset rams having a drink on their way home after being drafted out from the ewes through a gate

Once again you can see the importance of holding and covering while forcing. As Sid Cavanagh has often told me, 'A good dog must hold and force at the same time'. This was one of the things he liked so much about Barravore Jim.

Campaspe Joanne (Castleton Joe × Campaspe Speck) putting a mob of cows and calves over a creek swollen by recent rain

Most dogs rush in to force in one way or another, and then break back out to retrieve the resultant breakaways. The good dog doesn't allow any sheep to break away in the first place. Once stock have broken away once, they will try again. Dogs that over-head or don't cover, allow (or cause) stock to break behind them.

To put sheep across a creek, you can also try the same method as for putting them through a gateway. Work the mob along the creek bank and send the dog up the wing to ring the outside sheep around the lead and rush them into the water.

Another method, with well-trained dogs, is to split as many sheep off the main mob as you think you can handle, and put them across first by simply crowding them into the water with good holding/forcing dogs. You can ride your horse across in front of them which can help lead them across sometimes, but also gives the dogs something to force towards (particularly if they are inexperienced dogs, or lack the necessary drive to hold and force when taking stock away from you). Once they are across, put the rest over after them. If you can't manage this then you can catch a sheep (or even better a lamb) and carry it across, and tie it to something.

Stockwhips

If you haven't over-used the stockwhip its surprise value can rush the sheep into the water, if used at just the right time and in just the right place. This is not much use though if you are working whip-shy dogs.

I remember one incident taking a mob of cows and calves up a steep hill with two dogs, both of which were whip-shy, and I hadn't used the whip until nearly at the top. By then the cattle were pretty sick and tired of the long steep climb, and really began to baulk, and the dogs were pretty well knocked up. So out came the stockwhip and got things moving again, but when I looked around there wasn't a dog to be seen. I had to take the cows and calves the rest of the way on my own.

I haven't got much time for whip-shy dogs. However, other than breeding dogs with sound temperaments (which is the main factor), another trick can be of use in preventing whip-shyness. If you give the whip a crack or two each time you go out to feed a litter of pups, they will soon come running to the sound. This won't substitute for poor-tempered dogs, but it can help.

Handling weaners and lambs

Young sheep are often tricky to handle, and they soon show the good dogs from the poor ones. The lack of quality in dogs becomes very apparent when they can handle quiet older sheep but haven't a hope on young stock.

A white-hide stock whip and 70-year-old Australian stock saddle

The dog must be better in every respect to handle these types of sheep. The stock handle differently simply because they haven't been trained as yet, due to their limited handling. The principles for handling them are the same as for other stock, only everything has to be exaggerated.

When straightening or turning the mob, the dog or handler must go further to turn them. Older stock anticipate and straighten themselves up before the dog has got anywhere near them, whereas young stock don't – everything has to be spelled out to them in full.

Having said that, however, one of the most important points when working young stock is not to *over-head*. You must send the wing along and use it to ring around the lead, with the dog working on the shoulder of the mob, or at least working behind the shoulder of the lead sheep, to turn them. The same applies to any mob work, it is just more exaggerated when working young stock, and they are far less forgiving.

The good dog drives and balances sheep, and works them from behind, not from in front of them. Such a dog with great short cover and balance and hold together is a master at this sort of work.

A top covering dog is worth watching as it causes a mob to ring around in the required direction in this manner, moving towards the lead to turn it, and then back a bit to cover any that try to break behind it (often only a matter of a metre or so or even centimetres), but keeping the mob together and the outside sheep ringing up towards the lead.

When working older stock you can get away with poorly bred dogs (at a pinch) that are always wanting to head stock instead of driving and balancing them, but not with young stock. The mob will split up behind them. Young stock panic more readily and split off from the main mob.

Also any dog that lacks cover and break and width and hold together will not be able to handle young stock, because it will tend to cause them to split in every direction. Neither will a dog that breaks wide with too little provocation and doesn't block and hold, because the lambs will follow it and gain ground. Any turn tail will also become evident as the sheep attempt to break around and over the dog.

In my opinion about the most challenging *mob* work with sheep, to really test a dog (and its handler), is working a big mob of shorn merino lambs in rough country, and trying to put them somewhere they don't want to go. (A mob of stirry young cattle is a great test also.) I remember, as clear as if it was yesterday, one particular incident working a mob of about six or seven hundred merino lambs that had just been shorn and weaned, and had to be taken to a paddock some distance away, through a number of other paddocks.

I was on my own with one dog, Moorlands Georgie, and the country was hilly and rocky and broken, with a number of gates to negotiate right on the top of rises with the ground falling away from them in all directions. This really tested George out, and she showed her mettle that day.

(Then you get a certain type of trial handler complaining about sheep being too wild on a little cricket ground, and wanting them quietened down before they work them! If only they could try working merino lambs in rough country, or try sending their dog around a few shorn wethers running in their home country half-a-kilometre away up the side of a hill among the rock and the timber. Then they might start to understand what wild sheep are, and what a good dog is. Out in the hills you don't get to quieten the sheep down beforehand, or shear the wool from around their eyes, or put them in a good spot for an easy pick-up, or anything else.)

Working young cattle

It is sometimes said that cattle work like young sheep. This is only true when the cattle haven't been worked by good dogs. Young cattle, or unhandled or poorly handled cattle, do work like young unhandled sheep. If handled properly, they come to handle the same as a mob of older sheep.

The difference is that a dog's weakness becomes more obvious when working cattle than when working sheep, so the cattle often don't get trained to handle as well because they have learnt that they can walk all over the dogs. Good dogs make quiet cattle, as they do sheep, and make them quickly. The poorer the dog the longer it will take to quieten stock, or it may never occur.

Quietening stock

If you have a mob of sheep bolting away up a road, and leaving the tail of the mob behind, you can send your dog to block the lead and turn it back (and so correct them) a

Excuses are no use out here

couple of times, and they will generally settle down to a more sedate pace, because you have corrected them for running.

Remember that you and the dog are handling the sheep in the same way that you train a dog. Every action has an effect in modifying the way the stock behave. The dog is rewarding or correcting the stock depending on what it is doing, just as the handler does the dog. The better the dog's natural ability the more closely it follows sound principles of training in its dealings with stock. This is one reason why there is only one way a top dog can work.

A good handler makes a good dog, and a good dog well-handled makes good stock.

A dog with great short cover and balance will correct sheep and reward them almost instantly they do the right or wrong thing, whereas the lesser dog will correct them late and then reward them late, and so the sheep take much longer to become well behaved, if ever. Also the good dog doesn't give them any option other than the right one and so they go where they are required much more easily.

An over-heading dog that is always getting in front of sheep, and therefore correcting them for every move, is like a handler who is always down on a dog and can never be pleased. The dog will become sour and eventually give up trying, as will the sheep when handled by a dog constantly heading them or getting too far towards the head when it shouldn't.

In contrast, a calm driving dog leaves stock alone, and so rewards them when they are going in the right direction. Also a dog with distance (not wide working, but the

With a mob of recently weaned calves

ability not to be drawn by stock) will force when required but then back off when the sheep respond, and so reward them. The dog that bores into sheep all of the time never rewards the stock for doing what is required, and so they will become hard to handle, or doughy, depending on how much holding ability the dog has.

Paradoxically a pushy dog will, in the long run, make stock harder to move, and cause them to travel more slowly. The handler with this sort of dog then generally thinks that he needs a dog with more force.

With Cullens Kiara (Bailey's Harry × Cullens Gael)

At the opposite end of the scale, other stockmen think that the way to make quiet cattle is to keep right off them when they are running, and just allow them to go along at their own pace. Then, when the cattle settle down, the handler comes in closer. But what they are doing is rewarding the cattle for running by keeping away from them, and then punishing them for slowing down by coming in close to them.

In fact what the stockman should do is keep up with them, and keep a grip on them with good dogs when they are running (so that they gain no advantage by running), until they quieten, and *then* back off. Otherwise you reward them for going fast by keeping off them, and so they will do this every time they are worked until they slow down simply due to the distance they have travelled. If they do quieten down over time (weeks or months) with such handling, it is not because of being handled by keeping right back off them, but through the times when they are turned or blocked up and so on, and when being put through gates or into the yards.

It is the same as breaking in a timid, flighty young horse. You don't quieten it and develop trust in it by keeping right away from it. It may be quiet while you stay away, but when you get close you will see that it isn't quiet at all. You quieten it by getting in close and showing it that it has nothing to fear, and that jumping around and behaving in a frightened manner doesn't benefit it, but that relaxing and standing quietly does. The same applies, of course, to all animals.

The quickest method to quieten stock (say for example some stirry young cattle), is simply to use good dogs and block any attempts to run, and hold them in one place. Then once they settle down allow them to move off. Then any time they run send a couple of good dogs to block them. Leave them alone if they are going quietly, block them up if they run.

Correct what you don't want, reward what you do.

The well-known stockman of early last century, Henry G. Lamond, had this to say:

> The cattle are panting and blowing as they come on camp. It isn't that they've been bustled. It's only the softness of their condition telling its tale. Fast work with cattle is bad work. There are odd exceptions to this rule; but they're so few that they can be passed over. The steadier cattle are taken the better it is

Dust

for them. The quieter they are worked the more profit there is in them. And the final test of the whole show, and the reason of the cattle being there, is for that same profit! Galloping horses, swirling lariats, smashing whips, and those things are right in their proper place – in a story-book or on the pictures. If cattle go steady let 'em go steady; if they go steadier let 'em go steadier; and if they won't go steady then steady them by force, if need be. (Lamond 1931)

But you need good dogs. Good, strong, calm dogs make quiet cattle. The good dog should run around to block the cattle, keeping out off them, and stand its ground and only apply more force if they keep coming. The next time the dog goes around and stands quietly in front of them, they will stop. The dog that rushes around and chops in at them is not ideal.

The well-trained stockhorse 'ground ties' when the reins are dropped on the ground, while the stockman attends to a sheep etc.

(However before the cattle quieten down to some extent you will have to use the shoulder of the mob to ring around the lead to stop them).

So pushing and blocking stock can effectively quieten them, and keeping back off stock can encourage them to go faster and remain timid and flighty – the opposite to what most people do.

Every action has an equal and opposite reaction.

For example, just as keeping off stock will not quieten them effectively, so mollycoddling a timid, soft, or half-hearted dog will not improve it. Or if you put pressure on a dog to 'sit down' and it turns and goes away from the sheep, you shouldn't back off because you are worried it might lose interest, because you would be rewarding it for losing interest. You should keep at it until it lies down, then back right off and let it work.

The best and quickest way to quieten stock is with good dogs, well handled, worked from a horse. In most situations you cannot work stock (or dogs) properly from a bike or a vehicle.

Working sheep in yards

Real natural ability is as valuable in the yards as it is in the paddock. The top holding and driving dog with strong force and with great cover, that doesn't split sheep, with calm controlled bark and minimal eye, makes the best yard dog.

A top short-covering dog gets sheep moving by that side-to-side movement, where the dog lacking cover doesn't. It moves just back behind the shoulder of sheep to move them forwards, then moves forwards to control and turn that forward movement.

The top balancing/driving dog is in the right place, and applies force in the right place. Lesser dogs apply force in the wrong place, and often cause sheep to turn back or huddle together, etc. They also lack the cover and 'block and hold' and strength to hold while they force.

Sheep should not get back past a good yard dog. When forcing up to the draft, or into the shed, it is very common to see sheep constantly breaking back past a second-rate dog and the dog having to leave its forcing to go and retrieve them. Force without hold is not ideal.

A good rule in all stock work is to use the least amount of force necessary.

This allows the sheep room to flow, and to look where they are going. Too much force only causes them to jam up and to panic. They feel cornered, and resist being forced. Use the least force and the least noise possible. Use force only when required, then back off and let the sheep flow.

One point to remember about bark is that it tends to be 'indiscriminate force', in that it affects all of the sheep. Therefore if the lead sheep are just starting to go into the shearing shed or into the drafting race, and the dog barks at the back of the mob to keep the tail pushed up, the bark can correct the leaders and cause them to turn back.

In contrast, quiet holding and forcing dogs only affect the sheep they are directly working, and so they can be forcing and holding the tail tightly, without adversely influencing the lead sheep. Therefore, quiet force is capable of being used with more precision than bark.

Most Border Collies don't bark, which is a handicap, but many Kelpies bark for the wrong reasons and far too much, which is also a handicap.

Working sheep in the shed

When working sheep in the shed, high quality dogs make a big difference. A top covering dog, with strong force, is about the best shed dog you will find. And most of the time they are better off working on the ground and working silently, because silent force is more precise and directed.

A bit of bark is all right provided it is used judiciously, only at the right times. Used at the wrong time it generally only causes more work by piling sheep into a corner. Most dogs that bark naturally bark far too often, and therefore at the wrong times. Similarly, many natural backing dogs spend too much time on the backs when they would be more effective on the ground.

It is also interesting to note that a dog with good 'hold together', that doesn't split sheep or leave sheep behind, also makes a better shed dog as well as a better yard dog – they hold while they force. Second-rate dogs always have sheep breaking back behind them and so on.

Sheep awaiting shearing

With a good dog you shouldn't need to close the gate behind them; they should hold sheep so well that they never get back past them.

Moorlands Georgie was about the best shed dog I have seen, penning up, and she was a top covering, tight holding, big casting/mustering dog, without any bark. A barking dog may sound as though it is doing something (usually just getting on everyone's nerves), but the silent, high quality dog will get the job done in half the time with none of the fuss.

How many dogs to work

It often seems that those handlers with the most poorly trained dogs always work the most dogs at once. They would generally be better off working fewer dogs and having those under more control. These people tend to think that the reason they can't get stock through a gate or into the yards or around the yards etc., is because of a lack of enough force, and so they work more dogs.

The problem is usually not a lack of force, but misdirected force and a lack of precision.

However, when working off a horse you generally need to work more dogs than when working off a bike. Your dogs need to be better all-round – stronger, with better force, and better cast, and better distance, and better trained, and better stamina. This is one reason why better dogs were the rule in the past.

When you are on a bike the stock generally run from it, and so less force is needed, and fewer dogs are needed. On the bike most farmers tend to cast themselves out, whereas on the horse you tend to cast the dogs out more, and so the dog needs more casting ability. It also needs more ability all round because it is often working stock further away from you. It also needs more distance so it doesn't cause weak sheep to lie down and have to be carried home. Nowadays it isn't uncommon to see a certain type of so-called 'stockman' driving around with a trailer attached, to carry all those sheep home that their pushy dogs have caused to lie down.

If you have good, steady, calm dogs, with good drive, you can work a number of them together without getting into trouble. The better the dogs, the easier it is to work more of them together. But if you have half-mad heading dogs, or pushy dogs, you will have your hands full.

A good rule is to work the minimum number of dogs required. One dog worked with precision will do a better job than two dogs in the wrong place most of the time. But of course with bigger mobs, and rougher country, and hotter weather, and greater distances, more dogs are needed.

32

The dog in hard work

Some years ago I wondered just how much running about a dog does, in a day's work on a big property. So one morning I measured the distance I rode on a bike, in about three hours of light work. The bike travelled 35 kilometres, and the dog was on the ground travelling the whole time, either following the bike or working sheep. It did something more than 35 kilometres in three hours. Therefore I have no hesitation in saying that a dog – in a full day's work on a big property – can easily do 80 or more kilometres.

When dogs are used for this level of hard work, you need to pay attention particularly to three main areas. First, feeding them properly becomes more important. Second, looking after their feet is vital. But the most important area is the dog's breeding. I will discuss each of these areas in turn.

Feeding

When it comes to feeding working dogs, the most common error is having the dogs far too fat. When really working dogs hard, and particularly in hot weather, the slightest bit of excess weight will knock them about badly. You never see a fat marathon runner, nor one built like a wrestler.

A dog should have virtually no real fat cover, and it should just be possible to see its ribs (depending on how much hair it has). However, this is not achieved by

Campaspe Meg – an athletic Border Collie in good working condition

starvation because that would reduce the dog's muscle mass as well as its overall weight (although most people feed their dogs far more than is good for them). What is required is to remove the fat and develop muscle.

This is achieved by sound diet (including raw meat on the bone whenever it is available) and hard exercise. So although its ribs should just be visible, the dog should have plenty of hard muscle and its coat should be shiny and healthy (depending on the dog – some dogs just naturally have poor coats).

Care of feet

A dog, like a horse, is only as good as its feet. Sometimes you may come across a dog with inherently soft feet, but more often other factors are causing the trouble. The main factor in dogs having tough feet is constant work. The dogs mostly likely to get foot problems are those that are only worked occasionally, or are carried everywhere in a ute or on the back of the bike.

Young dogs are prone to foot trouble, before their pads have toughened up. Also, sitting around on a chain for a long time will allow the feet to go soft. If you then throw the dog into work on hard ground you are asking for trouble, particularly if the dog has good 'block and hold' and stops or turns quickly and so forth.

You can sometimes have problems in a wet winter with lots of lush wet feed about. This will soften the dog's feet up and then, if you work it on hard going and it stops or turns suddenly, the pads can tear.

White feet

A common myth is that white feet are softer than black feet. I have always found that to be complete nonsense. I have never been able to see any difference in the toughness of either.

Mike Donelan, in *The Australian Kelpie*, made this observation:

> Of course, the pure Dingo has white feet. Many stockmen will not buy a white footed dog. "Too soft" they say. My experience has been that this is yet another fallacy. Some of my white footed Kelpies have been tougher in the feet than any others. (Donelan 1984)

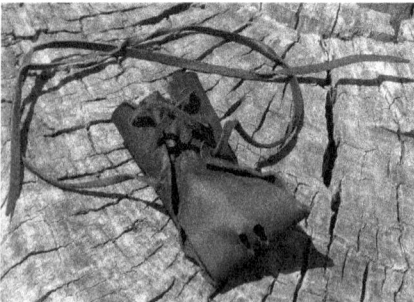

A dog-boot, useful for sore feet

Injuries to pads

If a dog has injured its pads, perhaps torn its stopper pad off when blocking a sheep on rock when its feet hadn't been toughened up to work beforehand, then it must be given time off and a chance to heal up.

The best remedy for torn pads is to apply honey to the pad each day. The dog will lick this off, but the honey doesn't have to be on very long to do its job. This is about the best thing to heal most wounds, including wounds on horses (and people also).

Once the pad has closed up and the skin grown back over, you can also dip the feet in a jar of a solution such as copper sulphate (bluestone) or zinc sulphate (be careful not to get any of either in your eyes, or the dog's eyes). These will harden the pads up. This is also a good idea if a dog has been out of hard work for some time.

If a dog has very dry, cracking pads then painting them with neat's-foot oil will help, but the diet might need looking at also.

Stamina

Stamina is vital in any working dog. Even when long, hard days are not the norm the dog with great stamina will do a better job, because it feels fitter. This is obvious on the three-sheep trial ground where the dog's performance obviously declines as it tires. The stamina of many three-sheep collies is fairly ordinary, but some are good.

Stamina is also becoming a significant problem in many lines of Kelpie, due to the breeding of dogs for yard trials that often move poorly and lack outright stamina. When you stop actively selecting for something, such as stamina or speed or agility, it rapidly declines, and so yard trials and breeders whose dogs are not actively involved in continuous stock work are doing damage in this regard. However, other lines of Kelpies still have good stamina.

But there are many aspects which combine to determine a dog's total 'stamina' – qualifications such as natural ability, conformation, size, water requirements, endurance of wind and energy, and particularly temperament.

Conformation

Conformation is one of those areas where you are best to avoid arbitrary rules, and let the end product speak for itself. However, a dog should have well-angulated shoulders and hindquarters, and plenty of leg under it.

The movement of the best-conformed dogs can be described in three ways: fast, smooth, and agile. The dog should move very flat as it goes, seemingly floating along, with its top-line hardly rising and falling with each stride. The less rocking it does, the better. Even a tall dog should look light on its feet. Many Kelpies bred from yard lines simply look big and flat-footed and sluggish, and are built more like wrestlers than athletes. Many also move in a loose fashion instead of tight and coordinated, and lack real speed.

The less weight a dog has to carry around the better, and so thickset, heavy-boned animals are undesirable. To use a human analogy a working dog should be something like a cross between a sprinter and a marathon runner.

Campaspe Sue (Cullens Jack × Cullens Kirribilli) at four months – an example of a well-conformed collie pup, fast and smooth; already she can outpace a grown sheep

A dog's tail going up in the air every time it stops or turns is often a sign of poor movement and poor conformation (and/or excitement).

Wind, energy, and water endurance

When it comes to outright stamina I believe that there are three areas to be considered. The first is the dog's 'wind endurance', which is its low requirements for, and ability to utilise, oxygen. The second is its 'energy endurance', which is how long it will go without getting tired and sluggish. The third is its 'water endurance', because some dogs are always looking for a drink while others never seem to need much water.

A dog that is 'good in its wind' is a dog that doesn't get overly puffed out, even in fast work. They seem to be able to work fast without getting out of breath. Temperament certainly plays a role, because a tense, excitable dog tends to get out of breath easily; you often see a certain type of Kelpie in the back of a ute with its tongue hanging out and puffed out just riding in the ute. But you get other Kelpies that never do much puffing at all.

Look for the pup that doesn't open its mouth very wide, and holds its tongue in its mouth with the tip curled upwards, even when it has been working hard. (This applies regardless of whether it is fit or not.) When you see dogs working with their mouths wide open, and the tongue lolling out the side (which is particularly noticeable in many UK dogs), you can be sure their endurance is ordinary.

As for 'energy endurance', some dogs can go all day without getting exhausted, and cope well with heat. Others knock up and are fairly useless for hard work; all right for trial work maybe, but not much good for serious work. You can sometimes get away with dogs of substandard stamina when you are working off a bike, because the dog is carted out to work and can ride between jobs and so have a spell, whereas on the horse the dog is always on the ground and travelling, and generally has more work to do.

I have owned and worked old dogs which you might be forgiven for thinking wouldn't have much stamina, but due to a combination of outright stamina of wind and energy, calm temperament and steadiness, and exceptional ability (not doing anything more than needed doing), could outwork most younger dogs of lesser breeding.

A dog should naturally have a good, hard, well-toned appearance (even if it isn't fit), as against many dogs that look soft and run to fat. (These soft, loose-looking sorts of dogs are also more prone to problems such as hip dysplasia. Also having pups too fat and growing too fast can exacerbate these types of inherited tendencies.)

And finally, 'water endurance'. Some dogs are always looking for water when working, while others never seem to, and when they do they only take a sip or two and then carry on.

Once again, you can see how many variables there are in the breeding of sheep dogs, which is what makes their breeding so challenging. There are so many things that must come together to make the 'top dog'. But when dogs are selected for the narrow requirements of restricted little dog trials many of these areas are neglected.

What we need are trials that ensure that trial handlers are breeding top practical working dogs, and working *for* the real sheep dog instead of against it, and then trial handlers will become a real asset to the working dog.

Temperament

As stated previously, temperament is a vital factor in a dog's 'stamina'. Nothing knocks a dog up more than a poor temperament – too excitable, tense, and busy. The good, steady, calm, relaxed dog is the one far more likely to still be with you at the end of a long, hot day. A dog should poke along at a walk or trot when that is all that is needed. Fast, busy dogs that are constantly back and forwards at a canter or a gallop are not much good at all, and neither are those that concentrate strongly when that isn't needed.

Temperament plays such a vital role in every area of a dog's work. No matter what you want a dog for, keep its temperament in the forefront of your considerations. Look for the very calm, relaxed dog. Avoid like the plague the tense, nervy, bull-at-a-gate variety.

There are two good indicators of a dog's temperament while it is working. One is the way it carries its tail. The tail should hang down more or less vertically, completely relaxed with no tension in it, even when in tight situations. A tail held high or tight or curled under or over in any situation is a sure sign of a tense, and probably weak, dog. The tail carriage is something that UK handlers set great store by, and rightly so.

The other indicator is the dog's breathing. Even when showing a lot of eye and concentration the dog should still hold its mouth open and breathe calmly. Many dogs, when the pressure comes on, clamp their mouths tight shut – again, a sure sign of tension and/or weakness, probably both.

Summary

In summary, in order to get the maximum stamina in any dog we need a combination of all the above traits. We need a dog that is athletic and well built, and capable of moving very fast and freely with ease and economy of effort.

If we can combine this with a calm relaxed attitude, in a dog that works steadily when possible and does no more running around than indicated (but as much as is needed), which isn't too intense when that is not required, with great stamina of wind and energy and water, then we will have a dog with the greatest possible stamina.

Part 6
DOG TRIALS

33

Sheep dog trials

Sheep dog trials have a powerful effect on the breeding of sheep dogs, because sooner or later most keen dog people take up the challenge. Unfortunately, once they do so their aim in breeding often changes from one of practical usefulness to one leaning more towards success at trials.

And because the type of dog suited to success at most trials is very different to that suited to practical work, the breeding suffers.

Apart from the fact that the trials themselves are so abysmal, many competitors have had very little real, solid, continuous experience at challenging stock work, and so have very little idea of what is required, which further compounds the problem. However, if the trials were sound this would cease to be an issue.

As stated earlier, trials create an *artificial form of necessity*. If this artificial necessity were to be brought into line with practical necessity, then trial handlers would become an asset to the working dog instead of a liability.

The difference in various types of dogs shows the power that trials have in altering their breeding. For example, the typical UK Border Collie is completely different from – and for practical work far superior to – the typical Australian Border Collie (but it is still far from perfect).

Unfortunately Australian three-sheep trials have resulted in destroying a good breed of sheep dogs, to the point where they are now mostly nothing more than show dogs.

Two of the Victorian Working Sheep Dog Association's stated aims are:

- to encourage and develop the standard of our working sheep dogs; and
- to take any action it thinks necessary to further promote the interests of sheep men and their working sheep dogs.

You will note that it is not to encourage the breeding of specialist trial dogs, nor to promote the interests of trial handlers and trial dogs, but sheep men and their working dogs. Most dog trial associations have similar stated aims, but unfortunately these stated purposes are hardly even given lip service today.

The current fad for making trials easier will only get worse as the make-up of trial competitors changes even further away from practical working stock people, and as the dogs continue to deteriorate.

Problems with present dog trials

Poorly designed trials result in a deterioration of the dogs being bred for them, which then necessitates lowering the trials' difficulty in order to accommodate poor dogs, which results in even more deterioration of the dog's ability. Therefore with poorly designed trials we have a downward snowballing effect.

On the other hand, well-designed trials would see an improvement in the quality of the dogs being bred for them, and so the level of difficulty could gradually be raised and thus result in even greater improvements in the dogs.

Changes to three-sheep trials such as the increasing of time limits from 12 minutes to 15 minutes, and of quietening sheep before the trial to make them easier to work, are highly detrimental (the working of relatively quiet sheep is also, in my opinion, often a problem with the national Kelpie utility trial).

Originally in Victoria the time limit was 15 minutes. However, many years ago the late Bill Marshall saw (even then) the deterioration that trials were causing in the breeding of dogs: 'They're breeding the dog out of them', he is quoted as saying. In an effort to do something about this he put a motion to decrease the time limits to 12 minutes, which was passed. This stood until a few years ago when it was changed back to allow 15 minute trials, in the push to cater for weak, mechanical dogs by hobby handlers.

Longer time limits allow handlers to 'drift' and humour sheep around the course with weak, mechanical dogs, whereas shorter time limits require a dog to take hold of its sheep and *put* them around the course. The shorter time limits encourage dogs with more outright holding ability and strength.

A good case in point when considering this trend to make trials easier is the Supreme Australian Championship. To my way of thinking, as the level of competition rises so too should the difficulty of the trials. This is what happens in the UK and elsewhere. Not so in Australia. The trials here are generally the same whether competing at a little local novice trial or the Australian Championships.

But the then decision makers of the Australian Sheepdog Worker's Association went one better. They developed a course for the Supreme Australian Championship which I consider to be the easiest three-sheep course I have seen. They have wider openings and obstacles, and panels and gates set on angles, and various other changes to make the course easier. They have long time limits, and also place great emphasis on quietening the sheep before the trial. Not overly bright, I wouldn't have thought. The challenge should rise as the level of competition rises, but here we have the highest level of competition in the country also being the easiest. I hope this will change in the future.

Drastic revision required

The design of three-sheep trials is now, more than ever, in need of drastic revision. They may have been partially adequate when most competitors were also stock people, and so their desire to breed trial dogs was tempered by the desire to breed practical workers. A number of such people bred some high quality dogs, and I think that the combination of hard practical work and three-sheep trials was one reason they did so.

Also many judges were practical stock people, and were looking for a practical working dog, and judged them accordingly. Not so nowadays. This is one reason why now more than ever the trials need drastic revision, but they were never satisfactory.

The three-sheep trial design itself is, and always has been, fundamentally flawed.

Many years ago James Moore commented on the poor design of Australian three-sheep trials:

> There are those who view with some misgiving, as I have been at pains to state elsewhere in this book, the tendency to run trials here merely for show purposes, and to bring about tricks which have no real bearing on the actual working life of the sheep-dog. (Moore 1929)

If we can align the requirements in a trial dog with those of a working dog, with more emphasis on the most important aspects (such as force, steadiness, drive, cast, natural holding ability, balance) and less on the not so important aspects (mad keen heading ability, weak dogs that quieten sheep) then the breeding will improve rapidly.

What we need are trials that force breeders to breed a practical working dog.

There is absolutely no reason why the two cannot go hand in hand – a good, enjoyable and highly competitive hobby that also results in high quality working dogs being bred. In the short term this will meet with a lot of opposition, because most established trial handlers realise that their dogs have no hope of competing in such a trial, and so they resist any change to the existing system.

In the long term such a change would be beneficial to all. Not only would the trials be far more interesting, but they would also result in breeding better dogs. At present it is close to impossible to buy a really high quality animal, Kelpie or collie, and very difficult to buy even a handy one. Some people may think that they are breeding good working dogs. But they are probably only working them on small places and are willing to put up with weakness and excitability, or dogs that split sheep or won't cast, because the dogs might be successful trial dogs with good handling. And quite probably they have never owned a truly good dog and don't know what is possible.

And don't think that yard dog trials are the way to go. They have serious shortcomings in the other direction. They allow dogs with very little holding ability, lacking break and cast and cover, that are often ratty and excitable, which don't get to the head, and which are often poor physically, to succeed.

Utility trials are currently the best type of trial in Australia. Unfortunately there are very few being run. If we could eliminate most three-sheep trials and yard dog trials, and concentrate on utility trials, then that would be a big step in the right direction. And if we could then increase the level of difficulty of the outside work of utility trials (and I certainly don't mean that we should turn the outside work into a three-sheep trial) we would see a great improvement in a short space of time, without any shadow of a doubt.

Unfortunately, the logic of many seems to be that if the trials have resulted in breeding poor dogs, then the solution is to make the trials easier. I fail to see how that is going to help anything, except perhaps give people a false sense of achievement when their poor excuse for a dog wins a ridiculously easy trial!

If we instigate trials that produce top, all-round dogs then there will be no need for specialist yard trials or specialist paddock trials, because specialist dogs will become a thing of the best-forgotten past in the face of exceptional all-round dogs that excel in *every* area.

34

Various trials – their limitations and benefits

In this chapter I will consider various types of sheep dog trials, and the way in which they result in certain faults or certain beneficial qualities being bred into the dogs used for them. It is worth repeating that sheep dog trials are enormously powerful tools in guiding the direction breeding takes, and *simply by altering the type of trial in common use we can alter the type of dog being bred, as sure as night follows day.*

I will consider the UK style dog trials, and the Australian three-sheep trials, yard-dog trials, and utility trials, and the effects each has.

UK trials

The UK trials, particularly the national and supreme courses, include casts of 400 or 800 yards; sometimes include recasting; include large portions of driving; include splitting sheep and sometimes working one; and penning. Also they often work a larger number of sheep out in the open than we do.

Generally they have resulted in a fairly high quality dog being bred. These dogs have great intelligence and can be trained very easily and readily controlled, with good heart, strong (partly due to the heavy nature of the sheep being worked, and partly due to the 'single' test holding one sheep from the others) with great scope, and that have natural driving ability and 'work their sheep from behind'.

However, in these trials the dogs are controlled every step of the way with far too much commanding. Also before they are trained such dogs will usually split sheep up and shoulder, and many bite badly. In contrast, the good dog must show 'breakaway break' when heading a breakaway, and good 'hold together' and width, *without any training*.

The reason such faults have developed is because their sheep are relatively quiet, and their trials have no real 'force and hold' obstacle. What is needed is an obstacle (as described later, or perhaps one even more severe due to the quiet nature of their sheep) that makes the dog really get in and force while the sheep are trying to break past and around it. This would show up their general lack of breakaway break and hold together, and their tendency to shoulder and bite, and hence the breeding would evolve to improve these areas.

In UK trials the dog is penalised for excess commanding on the outrun, which is good, but otherwise the dog is generally very much under tight command all the way around the course. One area that could be improved to help this is the draw, because this is a good area to get an idea of the dog's natural ability, provided it is allowed to draw the sheep back to the handler from a long distance *without any commands*.

The problem is the 'fetch gates' on the draw, through which the dog must draw the sheep to gain maximum points, and so the *handler controls the dog to achieve this*. If these 'fetch gates' were done away with, then the dog could be allowed to cast out and lift and draw the sheep *without a word of command*, and so its natural ability would be seen to advantage in these areas. There are enough other aspects that test the control that the handler has (and so test the dog's temperament) such as the large amount of driving work, so that removing these fetch gates would do no harm. In fact it would do a lot of good *by allowing the dog to be seen working without command*.

The driving aspect ensures that natural driving dogs are bred, with direct walk-in force. Many people in this country haven't experienced working stock with dogs with such direct force and drive, and don't realise how far superior they are to dogs that run around instead of forcing directly. The UK-bred dogs have great 'scope' and good distance, and are often willing to work very wide of their sheep due to this, and are often very fast in order to enable them to cover the great amount of ground that such wide work entails.

The penning at the UK trials is also fairly poor, with the handler doing so much of the work. If the handler were required to stand in one place, like we do here, and the dog made to do all of the work, then this would be better. This would also test the hold together aspect. Consider what C.B. Macpherson of Scotland had to say early last century in this regard: 'It has always seemed to me a tiresome sight as we do it – the man penning the sheep, the dog assisting' (in Moore 1929).

If these three changes were made, that is:

1   the fetch gates removed and the dog required to draw the sheep on his own resources *without command*;
2   the handler not allowed to assist the dog in penning; and
3   a real force and hold obstacle introduced,
    then the breed would improve to a very high standard.

They have a good foundation to build on (the best on average, in my opinion), both in their dogs and in their trials, but there is plenty of room for improvement.

Three-sheep trials

I consider three-sheep trials in Australia to be an absolute failure in encouraging the breeding of good working dogs. They mostly consist of working three sheep around a small cricket ground, and putting (or drifting) them through three fairly similar obstacles, and finally into a pen. One good aspect is that the handler is not allowed to assist the dog in any way, *which is as it should be*.

James Moore made the following comments in the 1920s about three-sheep trials:

> While in Australia there are a number of sheep-dog trial societies which have
> instituted open field trials, it must be pointed out that, since it is in agricultural
> show grounds that the majority of trials are held, limited in size and in scope, it
> has come about that a somewhat narrowing influence has crept in with the
> object of providing entertainment to the maximum number of the public. This
> restricted field of action certainly results in making dogs quick at hand and at
> negotiating obstacles, and sufficiently pliable to be moved quickly into
> position. Critics of this system, however, point out that a dog may be
> exceedingly clever in the restricted space of the trial ground without being able
> to carry these qualities into the wider field of the open pastureland in the
> course of his ordinary duties. (Moore 1929)

The type of dog that has evolved to compete at these trials is a dog mostly lacking in
any drive, and that is obsessed with getting in front of sheep and stopping them, because
that is the easiest dog to breed capable of putting three touchy sheep through such
obstacles. Also most are not much good at *naturally* working three sheep, even though
they may be successful at winning three-sheep trials with a lot of training.

Weak dogs are also bred, partly because they quieten sheep and partly because most
of the time there is little need for real strength or force in these trials. *As such, other traits
can be given priority when breeding.* Strong dogs tend to frighten the sheep, and are harder
to keep out wide (although not if they have the right ability, but breeding tends to sink to
the lowest level possible to get by). The dogs (as a generalisation) have little driving
ability, little mustering ability or scope, very little force, and are often hard to handle
because they are so mad keen to get to the head of sheep.

However, a top practical dog should be good enough to work three-sheep trials, *but
this is not the type of dog that the trials result in being bred.*

Boho Misty working a three-sheep trial (worked by Doug South after she was sold)

Basically three-sheep trials, as they stand, are a total failure at encouraging the breeding of practical dogs. They could be improved by the addition of a serious force and hold obstacle (as described later) in place of the first obstacle (the gap), so that a dog lacking in force and hold would not get past the first obstacle. If the cast was made a minimum of 400 metres also, and the judging method changed (as also explained later), and five sheep worked instead of three (because five require more cover and 'hold together') and a driving test introduced, then they would be a reasonable test. But it is hard to see how such drastic changes will ever be made on a wide scale.

Yard trials

Yard trials are also, in my opinion, counterproductive when it comes to improving the breeding of working dogs. Yes, yard work is *part* of a dog's work, *but a utility trial tests the yard work also*, and so I can see no place for pure yard trials in the context of encouraging the breeding of good dogs.

Unfortunately there are a great number of yard trials held throughout this country, and only a small number of utility trials. If that situation could be reversed the outlook would be much better.

The main requirements for dogs working in yard trials are that they are natural backing dogs, and that they have some bark and force. However, they generally lack much cover, break, casting ability, have poor temperaments, move badly and sometimes lack stamina, lack width or hold together, split sheep up, and don't get to the head. If they do have some of those traits it is usually because they have been bred as utility dogs rather than as specialist yard dogs.

There is nothing easier than to breed such a yard dog, or to train and handle one, and this is one reason why they have become so popular.

Utility trials

Utility trials, which are basically a yard trial combined with an easy paddock course, are better. The yard aspect tests that the dog has some bark and force, and will back, while the open work tests a bit more ability. *Unfortunately the open work is nowhere near challenging enough.*

If the outside work was made more demanding, with much longer casts (in my view anything less than about 400 metres hardly deserves to be called a cast), working five or six sheep instead of three, and the openings in the obstacles were made narrower (not the wings, but the openings) then they would be a reasonable trial. If driving away and the working of one sheep was incorporated, and a force and hold obstacle included, this would be good.

Summary

What the sheep dog breeds need is to be rid of all pure yard-dog trials and three-sheep trials, and to only concentrate on more severe utility or field trials. If such were the case then the standard of Australian dogs would rise, and rise rapidly.

35

Judging trials

It is not only the trials themselves that need drastic revision, but their method of judging also. First I will outline why this is so, and then describe a viable and realistic, and, for the most part, tried and proven, alternative.

I see three main problems with judging as it is currently practised (this includes three-sheep, yard, and utility trials, because all are judged by the same basic method). The first is that many judges are not practical stock people, and lack the knowledge necessary to do the job to a high standard, and they are not looking for good, practical dogs.

Second, the way in which most judge's judge is undesirable. They 'judge the sheep, and not the dog'. That is, they judge the position of the sheep. If the sheep step off course, points will come off; if they don't, they mostly won't.

Rather, much more emphasis should be placed on how well the dog works on its own initiative and instinct, rather than on technicalities.

To illustrate the above, a few years ago 'markers' were introduced to Victorian three-sheep trials to assist in this type of technical 'judging the sheep' mentality. In a three-sheep trial there is an imaginary corridor nine metres wide around the ground, from obstacle to obstacle, inside which the sheep are meant to be kept. In the past this was a rough guide. However, nowadays markers are placed to show exactly where this course is, and many judges line themselves up with these in order to see if a sheep puts a foot over the line.

So we have a very technical, but not very practical, method of judging. The dog can race to the head, then be sent back to the tail, and this is repeated over and over as a deliberate ploy by a successful handler working a mechanical dog, and yet if the sheep are inside the markers most judges will not deduct points for such mechanical work.

Yet a dog that has its sheep under good control *on its own*, balancing them in a straight line with hardly a word of command, and yet if they stray just centimetres over the line will be losing points. So you can see that common sense about the dog's ability has made way for this sort of technicality.

The good judge must also take into account the nature of the sheep being worked. With tough sheep they will be more lenient if the sheep are off course or get around an obstacle, providing that the dog shows great natural ability and is doing everything possible. Yet with quiet sheep the good judge will be much harder on lesser infringements, and penalise a poor mechanical dog even if it has an easy run. However, you then get the problem of the handler of that poor type of dog complaining that it didn't get a high enough score, when it had a supposedly 'good run'. So this type of judging has become rare, and the technical type has taken over. This is another reason that judging, and the trials themselves, need great revision.

The third main problem with current judging is the judging method itself, of simply removing points for faults, starting from 100 and deducting continuously from there. This is discussed further on.

We need trials (described in the next chapter) which themselves *prove the dog*, in order to eliminate these problems, but in conjunction with that we also need a better model of judging.

Cattle dog trial judging

One new method of judging has been implemented in cattle dog trials in Victoria, imported from America I believe. There is no judge as such, and the trial itself judges the dog, but in my opinion it is far from adequate. It has major drawbacks.

Yes, I agree that modern three-sheep trial methods of judging (which are also used in yard and utility trials) are poor, but the cattle-dog trial method is a case of having 'thrown the baby out with the bathwater'.

It works simply by allocating a certain number of points to each beast that goes through each obstacle. You only get one chance at putting the cattle through, and then move on to the next obstacle. So if working three cattle and all three go through the obstacle, you get 15 points (5 points for each beast). If you put all of the cattle through all of the obstacles, you get top points. If a number of dogs get top points, the winner is decided on time.

This removes the influence of the judge. However, it ignores many vital aspects such as the cast and lift, or how mechanical or natural the work is. There are no points allocated to the cast or lift whatsoever. The dog can cross the cast ten times on the way up the ground, ring around the cattle up the top before the handler's yells finally get it to bring the cattle down, and still go on and achieve the maximum possible score. Also the dog can throw in a dozen turn tails around the course, and not be penalised, and so forth.

If the cast, lift and draw were judged (other than the fetch obstacle on the draw, which I consider a bad idea), that would be a step in the right direction. For example by allocating 15 points to the cast, 5 to the lift, and 10 to the draw. The judge would allocate a score out of 15 for the cast, so a terrible cast would only get 0 or 1, a perfect cast would get 15, and the same for the lift and draw.

If you added another 15 points for the 'dog working on its own ability', it would be better again, but still fairly ordinary. So as a sound method of judging *in order to encourage the breeding of good dogs,* in my opinion it isn't worth considering.

Original intention of three-sheep judging

I believe that modern three-sheep trial judging is very different from that intended by the original instigators. In fact I am sure of it. I base this partly on the make-up of the scorecard, which doesn't make any sense as it is currently used. In fact many modern judges judge without recourse to the scorecard at all, and simply use a 'clicker counter' to keep track of how many points they have taken off, which bears out my argument.

Yet on the scorecard there are 12 points allocated to the cast, 5 for the lift, 8 for the draw, and 50 for general work. Nearly all judges completely ignore this scale of points, and simply take points off for faults as they see fit, starting from 100 and working backwards. Thus you can see that this allocation of points on the scorecard, as current judging exists, is completely ignored. Therefore, obviously the original instigators of the points system had something different in mind, or they would not have bothered allocating points in this way.

In the UK, trials are judged differently from Australian ones. A certain number of points are allocated to each section of the trial, and the dog given a score *out of that many for that section.* So if 20 points are allocated to the cast, the dog is scored out of 20 for the cast. A terrible cast, crossing the cast, might get 0 out of 20. *But the dog cannot lose more than that amount for that section.*

This, I am sure, is how the original instigators of the Victorian scoring system intended trials to be judged. It is slightly different to the UK system, because on the modern Victorian scorecard there are less points allocated to each obstacle than in the UK, but then 50 points are allocated to what is termed 'general work', in which is included 'force, steadiness, control of sheep and command of dog'. In the UK, general work is included in the points allocated to each section.

So in Victorian trials there are 25 points allocated in total to the four obstacles (gap 3; race 5; bridge 7; pen 10; total 25). If the dog pens its sheep it should therefore be entitled to at least 25 points for the obstacles, plus whatever score out of 25 it received for the cast, lift, and draw, plus its score for general work out of 50.

In the current system a dog can theoretically put up a first-class cast, lift and draw, and then make a mess of the rest of the trial, and lose enough points to give it a score of 0. Yet to my way of thinking, even though the dog messed up the rest of the trial it still deserves the points for the beautiful cast, lift and draw. This is not the case in the current system. I believe that what the original Victorian Working Sheep Dog Association councillors had in mind when they designed the scorecard was as follows:

Imagine judging a trial from start to finish. First the cast. You score the cast out of 12, keeping in mind that crossing the cast brings disqualification, so that a dog which sools straight up the middle of the ground just without crossing, and sools around the sheep up the top, might score 0 out of 12. (Crossing the cast doesn't bring disqualification in the UK, all points for the cast are simply lost; however, I am in favour of disqualification for crossing the cast *in arena trials* because if a dog crosses on a football ground it doesn't deserve to be out there; it would be a different matter on a much longer cast.)

Only a top, natural, wide cast, with the dog looking for its sheep and reading the point of balance at which to stop on its own, would score full points. Mechanical, man-made casts would be penalised hard.

Next the 5 points for the lift. If the dog stops on its own and lifts the sheep firmly but gently *without a word of command*, it might get the full 5 points. The worst possible lift with the dog sooling the sheep around might score 0. Once again a natural lift is the ideal; a lift that is controlled every step of the way by the handler, with the dog being backed off and pushed to this side and that, would get a very low score because it shows that the dog lacks the ability to be left to its own devices.

The same situation would apply to the draw, out of 8 points. A straight draw with the dog working on its own initiative, with the sheep well under control, is ideal.

So you then have a score out of 25 for the cast, lift and draw. Say, for example, the dog scored 20 out of 25. Now it continues around the course. Points come off the 50 points allocated to general work. The dog gains the allotted points for each obstacle completed, and loses the points for any not negotiated. If the dog gets the gap and the race, it would get 3 for the gap, and 5 for the race, thus a total of 8 points.

So it has 20 (out of a possible 25) for the cast, lift and draw, and 8 for completing two obstacles, thus a total of 28. *This is the minimum score that the dog can get for this run, plus any points allocated under general work.* Say the dog was given 10 for general work (out of a possible 50), then it would have a total score of 38 points.

The dog could not lose more than 50 points for general work – *otherwise what is the point of having 50 points allocated to general work if you take 100 off?*

This would mean that you would not have the situation of dogs which are 'a bit rough around the edges' but manage to put the sheep around the course, returning a score of 0, while a weak, pitiful, mechanical dog that only gets the first obstacle achieves a score of 30! I believe that this is a much better system and more in line with the intention of the original designers, though it is still not fully what they intended – it still has its problems. I would rather see the 50 points divided up, for example – 15 for force, 10 for steadiness, 10 for control of sheep, 5 for command of dog, and 10 for dog working on its own initiative as much as possible. In this way we would be more judging the dog, and not so much judging the handler and the sheep.

Another method

Another way of doing it is to go completely to the UK system of judging. This would mean that 'general work' as a separate category is done away with, and the points allocated to each section and to each obstacle of the trial increased to include all aspects of general work.

So you divide the 50 points now allocated to general work, between the cast, lift and draw and the four obstacles or sections. Thus you might have: cast 15; lift 5; draw 10; peg to completed gap 15; gap to completed race 17; race to completed bridge 18; bridge to competed pen 20.

If the worker fails to get to one or more sections they lose all points for those sections. So a worker only getting to the gap, but not putting the sheep through it, would lose all points for the race, bridge, and pen, thus they would lose 55 points, plus any lost in the cast, lift and draw, and any lost from the peg to the gap.

The score for each section would be decided in a different way to our current judging. Currently points are taken off for each fault, which results in a technical type of judging. I believe that the score for each section should be decided on a general impression of the dog's ability, and how well it is doing the job, taking into account force, steadiness, control of sheep and command of dog, and also how much the dog is working on its own initiative on its own natural ability, and allocating a score accordingly.

So you would simply look at the cast and think: 'What was that cast worth out of 15?' or 'What was that draw worth out of 10?' or 'What was the dog's work worth, taking the sheep from the peg to the gap, and putting them through, out of 15?' considering a really poor job being a zero, and natural perfection being full points.

It is also a much simpler, more transparent, more consistent, and more logical system of judging instead of the way that trials are now judged. Also spectators can more easily appreciate the manner in which the trial is being judged (although I don't consider spectator involvement as a necessary component of sheep dog trialling).

I will give one more point in support of my view that sheep dog trial judging is very different from that of years gone by. Once again I will quote James L. Moore:

> The total points awarded for the [old UK] trial are 55. Of these, five points are awarded for the 'run out', or as we call it in Australia, the 'cast'. Another five are awarded for the manner in which the dog 'lifts' his sheep ... Still another five are allowed for 'bringing' ... the next portion of the competition begins – that is to say, 'driving away' ... Ten points are awarded for proficiency in this department ... Ten points are allowed for 'shedding' ... The points allowed for 'penning' are five. Two more items need attention – one is 'style' ... 'Style' includes steadiness, concentration, and the general workmanlike attitude of the dog during the whole trial. Five points are allowed for this. Next comes one of the most important things of all – the dog's sense of command. For this 10 points are awarded.

and then a bit further on –

> The scale of points [in Australian trials] is given in a manner widely differing from the British standard. Under command, 20; steadiness, 10; general working, 20; negotiating obstacles, 30; style is divided into three sections – cast awarded eight points, approach five, command of sheep seven – total, 100. (Moore 1929)

You can see that the original principle of judging trials in Australia was the same as that originally used, and continuing to be used, in Britain, even though the allocations themselves were different. Points were allocated to each section (or trait, such as steadiness or force), and a score given out of that amount. I consider this to be the only sensible system of judging.

In this system you can 'weight' the more important aspects of the trial by allocating more points to them, and allocating less to the less important aspects, which you cannot do when points are simply deducted one after another each time a sheep puts a foot over a line. In fact this is the main reason such a system was instigated in the first place.

Nevertheless, regardless of the system of judging, the cardinal rule is to 'judge the dog, not the sheep'. Always look for the dog working on natural ability, showing the right make-up to truly be described as a 'working dog' and not just a show dog, and penalise weak and/or mechanical dogs accordingly, *hard,* so that such work is discouraged.

36

A true test

There should be a number of main requirements behind any sheep dog trial. First the trial must provide a challenge and an interest for the handlers, so they will enter. At the same time we should also design a trial that will provide the necessary 'artificial necessity' required to encourage the breeding of high calibre working dogs.

Another factor is that a trial should be a place where breeders can see a large number of dogs work, and see what abilities those dogs really possess, *by seeing them really tested in every area.*

Given the current poor quality of dogs, it is also important to design into the trial a method whereby handlers can compete at a lower level, until their dogs and handling improve. The trial outlined below meets this requirement.

The breeding of better dogs will be the main aim, and it will be the natural outcome if such a trial was widely adopted.

It will also be much more interesting to farmers and stockmen, as the type of dog needed and encouraged will be an exceptional, all-round, natural working dog. Instead of dogs getting worse and the trials having to be made easier, the dogs will improve and the trials can then be made even harder.

Test dogs' abilities, not simulate work conditions

There is absolutely no need to attempt to simulate actual working farm conditions in a dog trial, nor is it in any way practical to do so. We do not need to work a mob of sheep, or actually work the dog in yards, or shed, or truck.

What we need to do is to devise ways of *testing the dog's inherent traits* that relate to real working ability, in the setting of a dog trial on a smaller number of sheep.

The abilities I am talking about are those outlined in Part I 'Natural ability and selection'. *We need to test each of these traits.* If we can produce a trial that will result in the dog with the best inherent, natural, all-round ability winning (of course coupled with being well trained) then we will have done well.

And then in the final (or in a championship event) we may produce an even more severe test of the better dogs and handlers, to allow them to show even more skill.

So we require the following elements in such a dog trial:

- to test the dog's natural abilities
- to test the dog's temperament
- an improved system of judging to ensure fairness, simplicity, transparency, consistency, and ensuring that the results are as they should be.
- a course (and judging method) that doesn't allow handling to hide shortcomings.

I will consider each of these elements individually.

Test dogs' inherited natural ability

Each necessary ability of the working dog should be tested by this trial, *and the trial biased towards those that are more important.* Also tests must be devised that will ensure that dogs with bad faults (such as weakness or excitability or mad keen heading instinct or lack of holding ability or drive) will fail.

The major abilities to test are:

- cast (includes 'scope')
- 'strength' and 'force' (with 'hold' and 'cover')
- 'hold' (including 'square cover', 'width', 'break', 'hold together', 'freedom of movement')
- 'drive' (including 'balance' and 'distance')
- steadiness.

In order to successfully test and judge these elements we need to design a *course that will prove the dog,* and that will *not allow handling ability to hide the dog's shortcomings,* and also to judge the trial in such a way that handling ability that hides faults and shortcomings will be penalised. Therefore on the cast, lift and draw any commanding will be penalised, because the natural dog shouldn't need any commanding, and we want to see what the dog's natural instincts are.

And we have obstacles designed so that handlers cannot drift sheep through with weak dogs, but for which the dog must have plenty of hold and drive and balance and clean force to put them through. This will ensure that it has plenty of natural 'breakaway break' and hold together and cover and strength and so on.

Also each obstacle should test a different aspect. I can't see any value in having three or four very similar obstacles all testing basically the same things, as in three-sheep trials or the open work of utility trials.

Test dogs' temperament

We will also incorporate aspects into the trial that ensure that dogs with good temperaments are more likely to be successful. Driving sheep away from the handler is one particularly good test of a sound, calm, keen, controllable temperament (and of square break), as is cutting one sheep off. But so are most aspects of the trial, because temperament affects every area of a dog's work.

Method of judging

The method of judging will be similar to the one outlined earlier, in which each section of the trial is allocated so many points and a score awarded out of that allocation. That score will be based on a general judgement about the dog's performance rather than on technicalities. The judge would take careful note to penalise mechanical dogs with little natural ability and to award natural workers that show natural balance, and drive, and holding ability, and steady controlled work and force.

In order to allow all handlers to 'have a go', and to provide the maximum chance of showing their dogs off to the best effect, crossing the long cast would not bring disqualification but rather a reduction in the score if crossing was not indicated. Deliberate crossing will be a part of some sections.

Also handlers may choose to skip any sections of the trial if they feel that their dog is not up to them, but would lose all points for that section. This would mean that every handler gets the opportunity to show their dog's abilities to their best advantage, and to gain experience, even though they may not be capable of certain parts of the trial.

There would be no time limit as such, partly in order to allow handlers to skip certain sections of the trial. But if the judge thought that a dog was struggling to finish a certain section, and the judge had seen enough to give the dog a score for that section, they would simply ask the competitor to move to the next obstacle. Thus we wouldn't have the situation where a dog and handler remain at one obstacle for 10 minutes, either with the dog and sheep going round and round, or the dog and sheep 'camped' doing nothing much at all, and boring everyone to tears either way.

A true test – the trial course

The following trial design (see Figure 36.1) overcomes the limitations of the various trials in existence, and improves on them in many ways.

I haven't included any distances (other than for the cast), or sizes of obstacles, because initially the distances may be shorter and the obstacles wider until the dogs and handling improve, when they can then be made more severe.

Cast, lift, and draw

The first aspect of this trial is the cast, certainly not shorter than 400 metres, and preferably twice that length if possible, as the dogs improve. This cast should help ensure

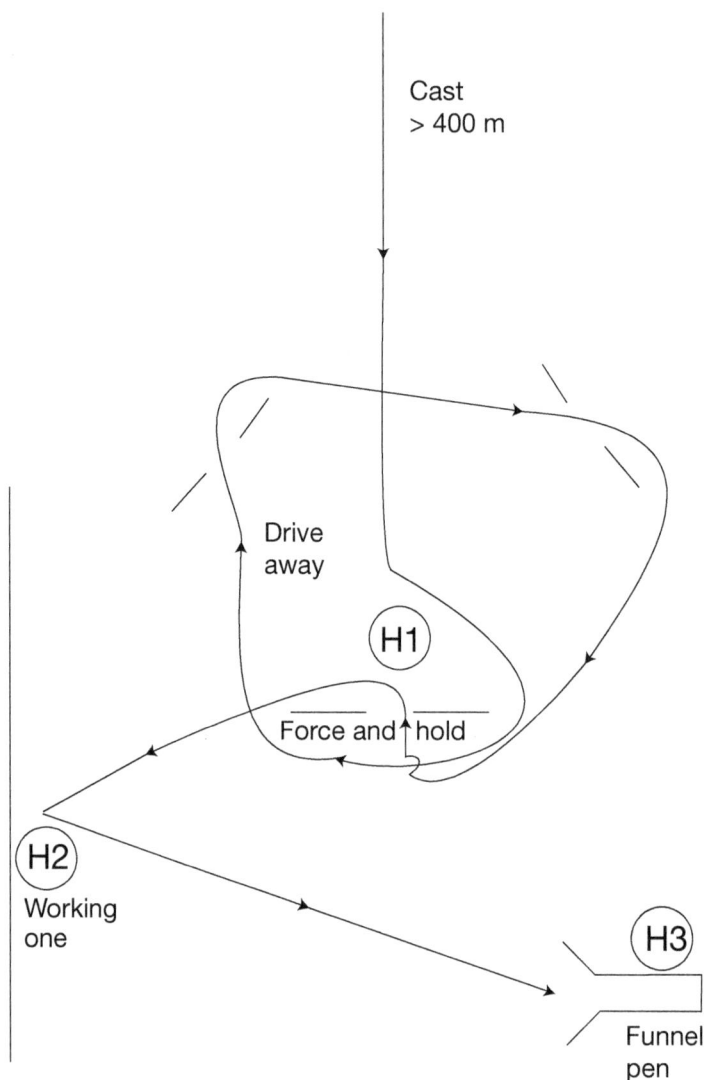

Figure 36.1 'A true test' trial course

that natural free-casting dogs with good 'scope' are bred, dogs that will look for their sheep and naturally lift them and draw them down the paddock.

The draw is one of the best indicators of a sheep dog's ability to drive and hold, and of its distance and balance and steadiness and calmness. However, you can't get a good idea of this ability on a small football ground. *You need a long cast to give a long draw.*

Any commanding on the cast, lift, or draw is penalised in order to reward the dog that does it naturally. In fact, any excess commanding in any part of the trial is penalised heavily, because we are looking for a natural worker. However, the dog will need to be well under control for the driving test, and this will test its temperament and 'controllability'.

Five sheep are the minimum number to be worked. Three sheep do not require a dog to have as much 'mob cover' as five, and working at least five sheep will encourage the breeding of dogs with a better idea of keeping sheep together. The sheep will *not* be quietened down before the trial. *We want a test, not a picnic.*

Driving

The sheep are now driven around below the 'force and hold' obstacle, and then driven away from the handler, through the first drive gates, across through the second, and then back down to the bottom side of the force and hold obstacle, ready to attempt that obstacle. The handler remains at the force and hold obstacle throughout.

Alternative drive test

An alternative to the drive in the open paddock through the drive gates is a 'fence drive' (see Figure 36.2). For example, the dog would drive the sheep away from the handler over to the side fence of the paddock, and then drive them up this fence. A number of drive gates (or pegs or stakes) would be placed out from the fence.

This has certain benefits over the open drive, in that the fence gives the dog a guide, and so it can be asked to drive the sheep more on its own initiative and less under control, and its purpose seen. This is a good test of temperament, and natural ability, and 'brains', and with merinos it may be easier to achieve than driving in the open.

Admittedly driving merinos is more difficult than with UK breeds of sheep, but it is such an important test that this must be overcome. Trials in the past in Australia featured driving, so there is no reason why they can't nowadays – if the dogs are good enough, and well enough trained. Also the long draw gives the dog the opportunity, if it is good enough, to get the sheep broken in to some extent so that it can then drive them well.

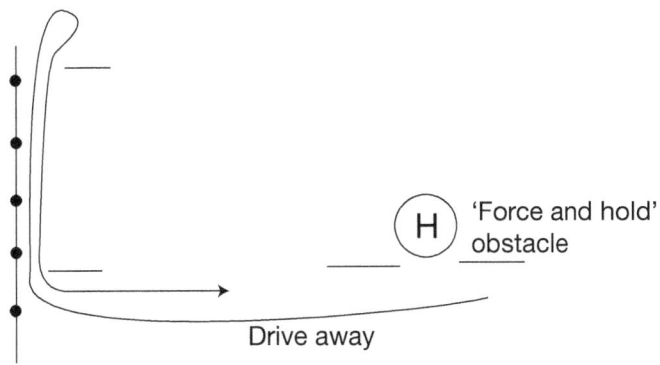

Figure 36.2 Alternative fence drive

'Force and hold' obstacle

There is not a single decent force and hold obstacle in any trial I have ever seen. Most trials test holding ability to some extent (although fairly minimally in yard dog trials), and most trials test force (although only minimally in three-sheep trials), *but none test outright natural holding ability combined with strong, clean force.*

C.B. Macpherson's comments early last century with regards to force tests in sheep dog trials are worth considering:

Most of our societies are very laudably trying to approximate, as nearly as may be, the work of their trials to the actual conditions of a shepherd's daily job, and are introducing "force" tests with that object. Having experimented at home on many kinds of tests, I have been disappointed in most, none of them being severe enough. (Macpherson in Moore 1929)

The requirements of a good force obstacle are that it provides sufficient resistance to the sheep going through, in order to prevent them being drifted through, and yet it doesn't simply cause the sheep to jam up in a narrow opening.

The obstacle I prefer, which meets these two requirements, is a simple gap obstacle with long wings, with an opening about 1.5 metres (five feet) wide (depending on how wild the sheep are, and so on) (see Figure 36.3). The handler stands directly behind this in order to provide the necessary resistance against the sheep coming through and to give the dog something to force and hold against.

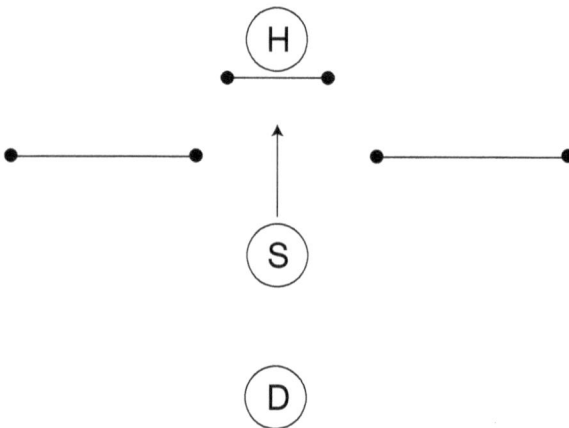

Figure 36.3 'Force and hold' gap obstacle

To complete this obstacle the dog will have to show plenty of cover (both mob and short) and real holding ability, combined with strong, controlled force. The dog that doesn't hold while it forces, but dives in to force or rushes towards the head as it forces, will never put the sheep through, nor will the dog that has the wrong type of eye or is weak, or just drifts sheep about, or bites.

This obstacle will ensure that the dog has plenty of break, because a dog that shoulders will do so when asked to force in this manner. It will also ensure that the dogs have plenty of short cover and block and hold (a big failing in most dogs).

On wild sheep this is about the best test of a dog's general purpose holding and forcing ability.

Even if the trial in its complete form is not instigated, but simply the force and hold obstacle introduced to all three-sheep and utility trials as the first obstacle, then that would be a big step in the right direction.

Alternative force and hold obstacle

An alternative obstacle is simply two long panels joining at an angle, with the handler standing behind the corner where the two panels join (see Figure 36.4). The dog is required to work the sheep into this corner, and to force them as tightly into the corner as it can. The judge gives the dog a score according to how well it does this.

Working one sheep

Once through the force and hold obstacle the handler proceeds to the 'working one' fence. The sheep are not required to be kept up on his right side (as in traditional three-sheep trials) as he is walking to the fence, rather the dog is simply left to bring the sheep along on its own and given a score (out of 5 points) for its work, without rushing the sheep around and without lagging behind, and doing so without commanding. It is allowed to do this on its own so that the judge, and the spectators, can see the dog's natural ability.

Figure 36.4 Alternative 'force and hold' obstacle

The fence to which the sheep are taken is either the paddock fence if that is not too far away, or a length of fence erected specially for the purpose. Against this length of fence one sheep is split off, and the dog required to show its ability to hold it off the others, and to work it away from them. *This is a fantastic test of the dog's ability in so many areas.* It is then recast around the other sheep to put them back together.

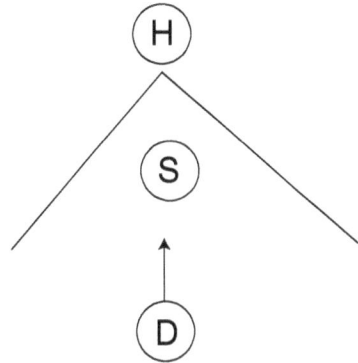

Yard course, or 'funnel' pen

There are two options for the final obstacle. One is to have a small yard course, to show the dog's barking and backing ability. The other is a 'funnel' pen. Some trials may have one or the other.

The funnel pen is another force and hold obstacle, different from the pen in common use. It has longish wings on both sides providing a funnel, and the pen gate set into one side as in the diagram below (Figure 36.5). The handler stands at the end of the gate, and closes it when the sheep are in the pen.

The funnel design allows the dog to show its 'block and hold' and force, and bark, to force the sheep along the straight, narrow section of the obstacle, before the gate is finally shut.

The difference between this, and the previous force and hold obstacle, is that the earlier obstacle was flatter. The dog had to show more ability to 'get around in front of the head' while holding, and more breakaway break and width and square cover, while in this obstacle the dog can show more block and hold and bark as the sheep try to get back past it.

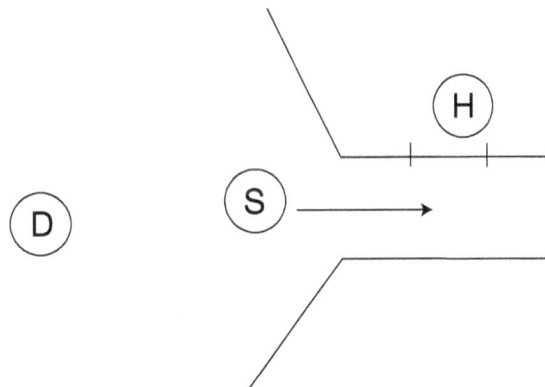

Figure 36.5 The 'funnel' pen

Macpherson also said:

> This leads me to a point on which I should greatly value expert opinion from the other side – at trials over here, as yet at least, a dog may not do that which his sense tells him is the obviously right thing to do to effect his object – he may not bark. (Macpherson in Moore 1929)

Sensible bark is valuable, and such an obstacle will test this, as will the judging method (5 points are allocated to barking, and, if a yard course is included, 15 points would be allocated to the yard course itself, which would include barking). The dog will be required to show its barking ability here, and asked to 'speak'.

Many dogs will not want to bark in such a pressure situation, because they have the wrong sort of eye, and many of the 'Jekyll and Hyde' type utility Kelpies will only bark if they rush in to force, and so they will not succeed either. Only the dog that barks sensibly while still holding and covering its sheep will succeed. Ideally it should bark *as it forces*, and not just stand back and bark, or look at the handler and bark.

Variations on the trial course

Other variations on this course are obviously possible. The four main areas that a good trial should contain, to encourage the breeding of good dogs, are:

- a long cast, *at least 400 metres*, which also gives a long draw
- driving away
- force and hold obstacle
- working one sheep
- back and bark test in yards, or funnel pen.

Specifics of judging

Each section of the trial is allocated so many points, and the dog scored out of this *on a general impression* of how well it handles the situation in question.

If a handler doesn't complete any section he loses full points for that section, but may skip that section and go on to the next.

Allocation of points:

- Cast 20
- Lift 10
- Draw 20
- Drive to first gate 6
- First gates 5 (full points are achieved if all sheep go through gates)
- Cross drive 6
- Second gates 5 (full points are achieved if all sheep go through gates)
- Back to force and hold obstacle 3
- Force and hold obstacle 25

- Walk to fence 5
- Working one sheep 20
- Walk to pen 5
- Funnel pen (or yard course) 15
- Bark 5
- Steadiness 10
  Total = 160 points

The handler cannot be worse off for attempting any obstacle or section, because they cannot lose any more points than are allocated to that section.

This method of judging allows certain aspects of a trial to be 'weighted' depending on their importance, that is more points are allocated to certain areas than to others, which you cannot do in the common method of judging modern Australian trials.

Summary

Sheep dog trials are very powerful tools in the breeding of sheep dogs, as shown by the different types of dogs that have resulted from different trials.

There is no reason at all why trials cannot be both a hobby and foster the breeding of exceptional dogs at the same time.

This trial is designed to encourage the breeding of top quality dogs, by testing all of the essential qualities in the real sheep dog so that breeders will select for these traits. If such a trial, including the features and judging that I have described, was instigated there would be very few dogs or handlers initially who could put up a good showing because most dogs about are so poor, and most handlers haven't had the experience driving sheep away or cutting one sheep off.

But everyone would be in the same boat initially. There would be some short-term pain, but the long-term benefits would be great. In a few years time everyone involved in dogs and stock would see the benefits.

I feel that if such a trial was instigated on a wide scale (and all three-sheep and yard trials eliminated), that in the space of 10 or 15 years we would see dogs being bred the like of which most people would not believe possible.

Initially when such trials were just getting started the gap of the 'force and hold' obstacle would be wider, and the funnel pen larger, and the 'drive' gates further apart and closer to the handler to make things easier. As time passed and the handling and the dogs improved, the gaps would be narrowed and the drive lengthened.

Also in the final of such a trial you might work more sheep to make the force and hold obstacle a greater challenge, and perhaps include some form of 'recasting' by casting first around half the sheep and then going back for the other half later.

One of the big benefits of this type of trial is that it gets around the problem of handlers and breeders and judges of limited experience not understanding what is required in a good dog. Such a trial will ensure that they will soon learn those points, and force them to breed good dogs even if initially they lack the knowledge to understand why such traits are essential.

If you are evaluating a dog, then consider how it would perform in this trial in your mind, when left to work on its own initiative as much as possible. This will give you a pretty good idea of its ability.

Consider particularly how well it would cast out and lift and draw the sheep back from the long cast without any commanding, and how easy it would be to get the dog to drive sheep away from the handler without looking to head them constantly, and how well it would hold and force five wild sheep through the 'force and hold' obstacle with the handler standing in the way, and how well it would work one sheep and prevent it from getting back to the others, or how well it would back and bark in the yard test.

You should quickly realise what a poor excuse for real working dogs most modern dogs really are.

What the *working* dog needs is a society with one aim, that of improving its breeding through the running of only such high quality trials, and through the maintenance of an elite stud book which registers dogs only on merit with no thought to breed or colour or anything else, but only exceptional *working* ability. Then inclusion in such a stud book would really mean something – a stud book based on instinct and natural ability, rather than on looks or ancestry alone – now there's a thought.

Breed is nothing; colour is nothing; type is nothing. Working ability is all that matters.

Afterword

Most modern dogs are of very poor quality, and I hope that this book has helped many people to realise the truth of this. I also hope that it has helped them to understand what a good dog really is, even if they have never experienced for themselves such a dog's instincts and abilities in hard, challenging work.

Making an accurate evaluation of the dogs you have, *and knowing what is possible and desirable in the long run*, is the first step towards breeding better dogs. Only by understanding how a dog should work, and the instincts and reflexes and temperament that it should possess, *and what is possible*, can this first step ever be taken.

The high calibre natural sheep dog is a marvellous thing, but it must be remembered that a sheep dog is a working dog. Anyone not breeding dogs for practical farm and stock work is not breeding working dogs. Rather they are only breeding show dogs, despite their protestations about 'trial dogs being good working dogs'. This also applies to those who put colour or type before working ability.

Poorly designed dog trials, and people breeding for them, have done – and continue to do – great damage to the working breeds. Only if they are redesigned can this state of affairs be reversed.

Buy (or breed) the best dogs you can, get them under good control to a sound, comprehensive commanding system, and allow them to work on their own initiative as much as possible. Then enjoy the magic of the good dogs.

I used to particularly enjoy working Moorlands Georgie just for the thrill of watching her in action. If I was in a suitable paddock of hills and gullies and rocks and timber, where I could stand on a high vantage point atop a ridge and get a good view, I would send her out on a mustering cast. Then I would watch as she curved around hillsides, dropped down into creeks or gullies and climbed back out the other side, broke out around rocky outcrops, clambered up steep rocky hillsides and disappeared out of sight over ridges, and shrank into the distance until she was only a black speck which I could hardly see, and often couldn't.

And all the while she would be starting some sheep in towards the centre of the paddock, then leaving them and turning and breaking back out, finding more and starting them on their way, and working her way right around the paddock in this manner, until all the sheep were moving in like the spokes of a giant wagon-wheel, picking their way over the ridges between the rocks, following the sheep tracks around the hillsides, or threading their way through the timber.

The author

Then I would see Georgie turning and retracing her steps, further consolidating the mob, until finally all the sheep were together, and then taking hold of them and working them back to me (which is a feat in itself in this class of country). And all without a word being said to her.

The gulf between the best dogs and the rest is enormous.

When breeding, concentrate particularly on a calm, bold temperament with plenty of 'heart', holding ability, force, drive, casting ability, and distance (which comes mostly from the drive and the calm temperament). And don't get preoccupied with fancy style because, as nice as it looks, it is mostly counterproductive. If you do this you won't go far wrong in breeding good, natural, brainy, obedient working dogs.

The rest is up to you. Good luck.

References

Bean CEW (1910) *On the wool track*. Harper Collins publishers, Moss Vale, NSW.

Cavanagh R (1990) *Australian sheepdogs – training and handling* Stocklight, Whittlesea.

Cramsie JB, Paton R, Reid SL, Gordon DJ, Macfarlane D, Potts HW and Mathews JW (no date) *Management and diseases of sheep in Australia*. Butterworth & Co. (Australia), Sydney.

Donelan M (1984) *The Australian kelpie*. Self-published by M Donelan, NSW.

Hagedoorn AL (1939) *Animal breeding*. Crosby Lockwood and Son Ltd, London.

Kelley RB (1942) *Sheep-dogs*. Angus and Robertson, Sydney.

Kelley RB (1960) *Principles and methods of animal breeding*. Angus and Robertson, Sydney.

Kohn A (1993) *Punished by rewards: the trouble with gold stars, incentive plans, A's, praise, and other bribes*. Houghton Mifflin, NY.

Lamond HG (1931) *Horns and hooves: handling stock in Australia*. Country Life, London Ltd.

Lush JL (1945) *Animal breeding plans*. Iowa State Coll. Press, Ames.

Mills AR, McIntyre WV and Herbert SF (1964) *A practical guide to handling dogs and stock*. A.H. & A.W. Reed, Sydney.

Moore JL (1929) *The canine king: The working sheep dog*. F Russell (Ed.), Standard Newspapers, Victoria.

Onstott K (1962) *The new art of breeding better dogs*. P Onstott (revised), Howell Book House, New York.

Pearse EH (no date) *Sheep and property management*. The Pastoral Review. Sydney

Index